Logistic Regression Examples Using the SAS® System

Version 6
First Edition

SAS Institute Inc.
SAS Campus Drive
Cary, NC 27513

The correct bibliographic citation for this manual is as follows: SAS Institute Inc., *Logistic Regression Examples Using the SAS® System, Version 6, First Edition*, Cary, NC: SAS Institute Inc., 1995. 163 pp.

Logistic Regression Examples Using the SAS® System, Version 6, First Edition

ISBN 1-55544-674-4

SAS Institute Inc., SAS Campus Drive, Cary, North Carolina 27513.

1st printing, March 1995

Doc P04, 25Jan95

Contents

Credits

Documentation

Design and Production	Design, Production, and Printing Services
Style Programming	Publications Technology Development
Technical Review	Jack J. Berry, Duane S. Hayes, Kristin R. Latour, David C. Schlotzhauer
Writing and Editing	James J. Ashton, Mara Beamish, Brent L. Cohen, Patricia Glasgow Moell, Josephine P. Pope

Acknowledgments

Frank Harrell of Duke University, Durham, NC, has been especially helpful in providing technical reviews of *Logistic Regression Examples Using the SAS® System.*

The final responsibility for the SAS System lies with SAS Institute alone. We hope you will always let us know your opinions about the SAS System and its documentation. It is through your participation that the progress of SAS software has been accomplished.

The Staff of SAS Institute Inc.

Introduction: The Example Data Sets

Featured Tools:

- DATA step
- PRINT procedure

This introduction describes data sets used for the examples in this book. The examples in this chapter show how to create a SAS data set for the raw data and print the data set.

The BRANDS Data Set

The brand choice data are hypothetical data from a brand choice study. In this study, 100 subjects choose one preferred brand from a set of five brands arranged in eight different brand/price combinations. For four of the brands, the price is either $5.99 or $3.99. For the fifth brand, the price is held constant at $4.99. The brand choice data set contains 10 variables: P1-P5 contain the brand prices; F1-F5 contain the frequencies with which each brand is chosen.

Program

Create the BRANDS data set.

```
data brands;
   input p1-p5 f1-f5;
   datalines;
5.99 5.99 5.99 5.99 4.99  12 19 22 33 14
5.99 5.99 3.99 3.99 4.99  34 26  8 27  5
5.99 3.99 5.99 3.99 4.99  13 37 15 27  8
5.99 3.99 3.99 5.99 4.99  49  1  9 37  4
3.99 5.99 5.99 3.99 4.99  31 12  6 18 33
3.99 5.99 3.99 5.99 4.99   4 29 16 42  9
3.99 3.99 5.99 5.99 4.99  37 10  5 35 13
3.99 3.99 3.99 3.99 4.99  16 14  5 51 14
;
```

Print the full data set.

```
proc print data=brands;
   title 'Brand Choice Data';
run;
```

Output 1
BRANDS Data Set

Brand Choice Data

OBS	P1	P2	P3	P4	P5	F1	F2	F3	F4	F5
1	5.99	5.99	5.99	5.99	4.99	12	19	22	33	14
2	5.99	5.99	3.99	3.99	4.99	34	26	8	27	5
3	5.99	3.99	5.99	3.99	4.99	13	37	15	27	8
4	5.99	3.99	3.99	5.99	4.99	49	1	9	37	4
5	3.99	5.99	5.99	3.99	4.99	31	12	6	18	33
6	3.99	5.99	3.99	5.99	4.99	4	29	16	42	9
7	3.99	3.99	5.99	5.99	4.99	37	10	5	35	13
8	3.99	3.99	3.99	3.99	4.99	16	14	5	51	14

The CHOCS Data Set

The chocolate candy data are hypothetical data from a discrete choice study in which 10 subjects are presented with eight different chocolate candies. The subjects choose one preferred candy from among the eight different types of candies. The eight candies consist of the 2^3 combinations of dark or milk chocolate, soft or hard center, and nuts or no nuts. The data set contains five variables:

SUBJ — an identification variable for each subject

CHOOSE — the response variable (0=not chosen, 1=chosen). Only one observation for each of the eight choice sets has a value of 1 for CHOOSE; the other seven have values of 0.

DARK — (0=milk chocolate, 1=dark chocolate)

SOFT — (0=hard center, 1=soft center)

NUTS — (0=no nuts, 1=nuts).

Program

Create the CHOCS data set.

```
data chocs;
   input subj choose dark soft nuts @@;
   datalines;
 1 0 0 0 0    1 0 0 0 1   1 0 0 1 0    1 0 0 1 1
 1 1 1 0 0    1 0 1 0 1   1 0 1 1 0    1 0 1 1 1
 2 0 0 0 0    2 0 0 0 1   2 0 0 1 0    2 0 0 1 1
 2 0 1 0 0    2 1 1 0 1   2 0 1 1 0    2 0 1 1 1
 3 0 0 0 0    3 0 0 0 1   3 0 0 1 0    3 0 0 1 1
 3 0 1 0 0    3 0 1 0 1   3 1 1 1 0    3 0 1 1 1
 4 0 0 0 0    4 0 0 0 1   4 0 0 1 0    4 0 0 1 1
 4 1 1 0 0    4 0 1 0 1   4 0 1 1 0    4 0 1 1 1
 5 0 0 0 0    5 1 0 0 1   5 0 0 1 0    5 0 0 1 1
 5 0 1 0 0    5 0 1 0 1   5 0 1 1 0    5 0 1 1 1
 6 0 0 0 0    6 0 0 0 1   6 0 0 1 0    6 0 0 1 1
 6 0 1 0 0    6 1 1 0 1   6 0 1 1 0    6 0 1 1 1
 7 0 0 0 0    7 1 0 0 1   7 0 0 1 0    7 0 0 1 1
 7 0 1 0 0    7 0 1 0 1   7 0 1 1 0    7 0 1 1 1
 8 0 0 0 0    8 0 0 0 1   8 0 0 1 0    8 0 0 1 1
 8 0 1 0 0    8 1 1 0 1   8 0 1 1 0    8 0 1 1 1
 9 0 0 0 0    9 0 0 0 1   9 0 0 1 0    9 0 0 1 1
 9 0 1 0 0    9 1 1 0 1   9 0 1 1 0    9 0 1 1 1
10 0 0 0 0   10 0 0 0 1  10 0 0 1 0   10 0 0 1 1
10 0 1 0 0   10 1 1 0 1  10 0 1 1 0   10 0 1 1 1
;
```

Print the first 10 observations.

```
proc print data=chocs(obs=10);
   title 'Chocolate Candy Data';
run;
```

Output 2
First 10 Observations of CHOCS Data Set

Chocolate Candy Data

OBS	SUBJ	CHOOSE	DARK	SOFT	NUTS
1	1	0	0	0	0
2	1	0	0	0	1
3	1	0	0	1	0
4	1	0	0	1	1
5	1	1	1	0	0
6	1	0	1	0	1
7	1	0	1	1	0
8	1	0	1	1	1
9	2	0	0	0	0
10	2	0	0	0	1

The DIABETES Data Set

Reaven and Miller (1979) collected data from 145 nonobese adults who were diagnosed as being subclinical (chemical) diabetics, overt diabetics, and normals. The study investigated the relationship between various blood chemistry measures and diabetic status. The data set DIABETES appears in Appendix A2.6 of Friendly (1991) and contains these variables:

PATIENT — patient number

RELWT — relative weight, expressed as a ratio of actual weight to expected weight, given the person's height

GLUFAST — fasting plasma glucose

GLUTEST — test plasma glucose, a measure of glucose intolerance

INSTEST — plasma insulin during test, a measure of insulin response to oral glucose

SSPG — steady state plasma glucose, a measure of insulin resistance

GROUP — clinical group (3=overt diabetic, 2=chemical diabetic, 1=normal).

Program

Create a format for the three types of diabetics.

```
proc format;
   value gp 3='(3) Overt Diabetic ' 2='(2) Chem. Diabetic' 1='(1) Normal';
run;
```

Create the DIABETES data set.

```
data diabetes;
   input patient relwt glufast glutest instest sspg group;
   label relwt   = 'Relative weight'
         glufast = 'Fasting Plasma Glucose'
         glutest = 'Test Plasma Glucose'
         instest = 'Plasma Insulin during Test'
         sspg    = 'Steady State Plasma Glucose'
         group   = 'Clinical Group';
   datalines;
   1  0.81  80  356 124   55  1
   2  0.95  97  289 117   76  1
   3  0.94 105  319 143  105  1
   4  1.04  90  356 199  108  1
   5  1.00  90  323 240  143  1
   6  0.76  86  381 157  165  1
   7  0.91 100  350 221  119  1
   8  1.10  85  301 186  105  1
   9  0.99  97  379 142   98  1
  10  0.78  97  296 131   94  1
  11  0.90  91  353 221   53  1
  12  0.73  87  306 178   66  1
  13  0.96  78  290 136  142  1
  14  0.84  90  371 200   93  1
  15  0.74  86  312 208   68  1
  16  0.98  80  393 202  102  1
  17  1.10  90  364 152   76  1
  18  0.85  99  359 185   37  1
```

```
19  0.83  85  296 116   60  1
20  0.93  90  345 123   50  1
21  0.95  90  378 136   47  1
22  0.74  88  304 134   50  1
23  0.95  95  347 184   91  1
24  0.97  90  327 192  124  1
25  0.72  92  386 279   74  1
26  1.11  74  365 228  235  1
27  1.20  98  365 145  158  1
28  1.13 100  352 172  140  1
29  1.00  86  325 179  145  1
30  0.78  98  321 222   99  1
31  1.00  70  360 134   90  1
32  1.00  99  336 143  105  1
33  0.71  75  352 169   32  1
34  0.76  90  353 263  165  1
35  0.89  85  373 174   78  1
36  0.88  99  376 134   80  1
37  1.17 100  367 182   54  1
38  0.85  78  335 241  175  1
39  0.97 106  396 128   80  1
40  1.00  98  277 222  186  1
41  1.00 102  378 165  117  1
42  0.89  90  360 282  160  1
43  0.98  94  291  94   71  1
44  0.78  80  269 121   29  1
45  0.74  93  318  73   42  1
46  0.91  86  328 106   56  1
47  0.95  85  334 118  122  1
48  0.95  96  356 112   73  1
49  1.03  88  291 157  122  1
50  0.87  87  360 292  128  1
51  0.87  94  313 200  233  1
52  1.17  93  306 220  132  1
53  0.83  86  319 144  138  1
54  0.82  86  349 109   83  1
55  0.86  96  332 151  109  1
56  1.01  86  323 158   96  1
57  0.88  89  323  73   52  1
58  0.75  83  351  81   42  1
59  0.99  98  478 151  122  2
60  1.12 100  398 122  176  1
61  1.09 110  426 117  118  1
62  1.02  88  439 208  244  2
63  1.19 100  429 201  194  2
64  1.06  80  333 131  136  1
65  1.20  89  472 162  257  2
66  1.05  91  436 148  167  2
67  1.18  96  418 130  153  1
68  1.01  95  391 137  248  1
69  0.91  82  390 375  273  1
70  0.81  84  416 146   80  1
71  1.10  90  413 344  270  2
72  1.03 100  385 192  180  1
73  0.97  86  393 115   85  1
```

74	0.96	93	376	195	106	1
75	1.10	107	403	267	254	1
76	1.07	112	414	281	119	1
77	1.08	94	426	213	177	2
78	0.95	93	364	156	159	1
79	0.74	93	391	221	103	1
80	0.84	90	356	199	59	1
81	0.89	99	398	76	108	1
82	1.11	93	393	490	259	1
83	1.19	85	425	143	204	2
84	1.18	89	318	73	220	1
85	1.06	96	465	237	111	2
86	0.95	111	558	748	122	2
87	1.06	107	503	320	253	2
88	0.98	114	540	188	211	2
89	1.16	101	469	607	271	2
90	1.18	108	486	297	220	2
91	1.20	112	568	232	276	2
92	1.08	105	527	480	233	2
93	0.91	103	537	622	264	2
94	1.03	99	466	287	231	2
95	1.09	102	599	266	268	2
96	1.05	110	477	124	60	2
97	1.20	102	472	297	272	2
98	1.05	96	456	326	235	2
99	1.10	95	517	564	206	2
100	1.12	112	503	408	300	2
101	0.96	110	522	325	286	2
102	1.13	92	476	433	226	2
103	1.07	104	472	180	239	2
104	1.10	75	455	392	242	2
105	0.94	92	442	109	157	2
106	1.12	92	541	313	267	2
107	0.88	92	580	132	155	2
108	0.93	93	472	285	194	2
109	1.16	112	562	139	198	2
110	0.94	88	423	212	156	2
111	0.91	114	643	155	100	2
112	0.83	103	533	120	135	2
113	0.92	300	1468	28	455	3
114	0.86	303	1487	23	327	3
115	0.85	125	714	232	279	3
116	0.83	280	1470	54	382	3
117	0.85	216	1113	81	378	3
118	1.06	190	972	87	374	3
119	1.06	151	854	76	260	3
120	0.92	303	1364	42	346	3
121	1.20	173	832	102	319	3
122	1.04	203	967	138	351	3
123	1.16	195	920	160	357	3
124	1.08	140	613	131	248	3
125	0.95	151	857	145	324	3
126	0.86	275	1373	45	300	3
127	0.90	260	1133	118	300	3
128	0.97	149	849	159	310	3

```
129  1.16 233 1183  73  458  3
130  1.12 146  847 103  339  3
131  1.07 124  538 460  320  3
132  0.93 213 1001  42  297  3
133  0.85 330 1520  13  303  3
134  0.81 123  557 130  152  3
135  0.98 130  670  44  167  3
136  1.01 120  636 314  220  3
137  1.19 138  741 219  209  3
138  1.04 188  958 100  351  3
139  1.06 339 1354  10  450  3
140  1.03 265 1263  83  413  3
141  1.05 353 1428  41  480  3
142  0.91 180  923  77  150  3
143  0.90 213 1025  29  209  3
144  1.11 328 1246 124  442  3
145  0.74 346 1568  15  253  3
;
```

Print the first 10 observations.

```
proc print data=diabetes (obs=10);
   title 'Diabetes Data';
run;
```

Output 3
First 10 Observations of DIABETES Data Set

Diabetes Data

OBS	PATIENT	RELWT	GLUFAST	GLUTEST	INSTEST	SSPG	GROUP
1	1	0.81	80	356	124	55	1
2	2	0.95	97	289	117	76	1
3	3	0.94	105	319	143	105	1
4	4	1.04	90	356	199	108	1
5	5	1.00	90	323	240	143	1
6	6	0.76	86	381	157	165	1
7	7	0.91	100	350	221	119	1
8	8	1.10	85	301	186	105	1
9	9	0.99	97	379	142	98	1
10	10	0.78	97	296	131	94	1

The ESR Data Set

The values of plasma fibrinogen and gamma-globulin are measured (in gm/liter) for a sample of 32 people and are used as explanatory variables to predict whether or not the individuals are healthy. A healthy individual should have an erythrocyte sedimentation rate (ESR) of less than 20 mm/hour. Because the actual value of the ESR is relatively unimportant, the response variable only denotes whether the individual is healthy or unhealthy. The value of the RESPONSE variable is coded as 0 for healthy people (ESR < 20), and as 1 for unhealthy people (ESR >= 20). The ESR data appear in Collett (1991)* and contain the following variables:

ID	an identification variable
FIBRIN	level of plasma fibrinogen (gm/liter)

* Reprinted by permission of the publisher.

GLOBULIN level of gamma-globulin (gm/liter)

RESPONSE a binary response variable (0=ESR<20, 1=ESR>=20).

Program

Create the ESR data set.

```
data esr;
   input id fibrin globulin response @@;
   datalines;
1  2.52 38 0   2  2.56 31 0   3  2.19 33 0   4  2.18 31 0
5  3.41 37 0   6  2.46 36 0   7  3.22 38 0   8  2.21 37 0
9  3.15 39 0   10 2.60 41 0   11 2.29 36 0   12 2.35 29 0
13 5.06 37 1   14 3.34 32 1   15 2.38 37 1   16 3.15 36 0
17 3.53 46 1   18 2.68 34 0   19 2.60 38 0   20 2.23 37 0
21 2.88 30 0   22 2.65 46 0   23 2.09 44 1   24 2.28 36 0
25 2.67 39 0   26 2.29 31 0   27 2.15 31 0   28 2.54 28 0
29 3.93 32 1   30 3.34 30 0   31 2.99 36 0   32 3.32 35 0
;
```

Print the first five observations.

```
proc print data=esr(obs=5);
   title 'ESR Data';
run;
```

Output 4
First Five Observations of ESR Data Set

```
                    ESR Data

   OBS    ID    FIBRIN    GLOBULIN    RESPONSE

    1      1     2.52        38           0
    2      2     2.56        31           0
    3      3     2.19        33           0
    4      4     2.18        31           0
    5      5     3.41        37           0
```

The MATCH_11 Data Set

A large study in 1986 at Baystate Medical Center, Springfield, Massachusetts, investigated risk factors for low birth weight babies, with low birth weight being defined as less than 2500 grams. Determining risk factors for low birth weight babies is important because infant mortality rates and birth defect rates are high for these babies. Risk factors that were investigated were race, smoking status during pregnancy, history of hypertension, history of premature labor, presence of uterine irritability, and mother's weight at last menstrual period.

In the study, each mother who gave birth to a low birth weight baby was age matched with a randomly selected control mother (that is, a mother of the same age who gave birth to a normal weight baby). The MATCH_11 data set appears in Appendix 3 of Hosmer and Lemeshow (1989) and contains 56 pairs of mothers matched on age and contains these variables:

PAIR identification variable for each matched pair

LOW low birth weight indicator (0=birth weight >= 2500 g, 1=birth weight < 2500 g)

AGE age of mother in years

LWT	weight in pounds at last menstrual period
RACE	race of mother (1=white, 2=black, 3=other)
SMOKE	smoking status during pregnancy (1=yes, 0=no)
PTD	history of premature labor (1=yes, 0=no)
HT	history of hypertension (1=yes, 0=no)
UI	presence of uterine irritability (1=yes, 0=no).

Program

Create the MATCH_11 data set.

```
data match_11;
   input pair low age lwt race smoke ptd ht ui @@;
   datalines;
1  0 14 135 1 0 0 0 0     1  1 14 101 3 1 1 0 0
2  0 15  98 2 0 0 0 0     2  1 15 115 3 0 0 0 1
3  0 16  95 3 0 0 0 0     3  1 16 130 3 0 0 0 0
4  0 17 103 3 0 0 0 0     4  1 17 130 3 1 1 0 1
5  0 17 122 1 1 0 0 0     5  1 17 110 1 1 0 0 0
6  0 17 113 2 0 0 0 0     6  1 17 120 1 1 0 0 0
7  0 17 113 2 0 0 0 0     7  1 17 120 2 0 0 0 0
8  0 17 119 3 0 0 0 0     8  1 17 142 2 0 0 1 0
9  0 18 100 1 1 0 0 0     9  1 18 148 3 0 0 0 0
10 0 18  90 1 1 0 0 1     10 1 18 110 2 1 1 0 0
11 0 19 150 3 0 0 0 0     11 1 19  91 1 1 1 0 1
12 0 19 115 3 0 0 0 0     12 1 19 102 1 0 0 0 0
13 0 19 235 1 1 0 1 0     13 1 19 112 1 1 0 0 1
14 0 20 120 3 0 0 0 1     14 1 20 150 1 1 0 0 0
15 0 20 103 3 0 0 0 0     15 1 20 125 3 0 0 0 1
16 0 20 169 3 0 1 0 1     16 1 20 120 2 1 0 0 0
17 0 20 141 1 0 1 0 1     17 1 20  80 3 1 0 0 1
18 0 20 121 2 1 0 0 0     18 1 20 109 3 0 0 0 0
19 0 20 127 3 0 0 0 0     19 1 20 121 1 1 1 0 1
20 0 20 120 3 0 0 0 0     20 1 20 122 2 1 0 0 0
21 0 20 158 1 0 0 0 0     21 1 20 105 3 0 0 0 0
22 0 21 108 1 1 0 0 1     22 1 21 165 1 1 0 1 0
23 0 21 124 3 0 0 0 0     23 1 21 200 2 0 0 0 0
24 0 21 185 2 1 0 0 0     24 1 21 103 3 0 0 0 0
25 0 21 160 1 0 0 0 0     25 1 21 100 3 0 1 0 0
26 0 21 115 1 0 0 0 0     26 1 21 130 1 1 0 1 0
27 0 22  95 3 0 0 1 0     27 1 22 130 1 1 0 0 0
28 0 22 158 2 0 1 0 0     28 1 22 130 1 1 1 0 1
29 0 23 130 2 0 0 0 0     29 1 23  97 3 0 0 0 1
30 0 23 128 3 0 0 0 0     30 1 23 187 2 1 0 0 0
31 0 23 119 3 0 0 0 0     31 1 23 120 3 0 0 0 0
32 0 23 115 3 1 0 0 0     32 1 23 110 1 1 1 0 0
33 0 23 190 1 0 0 0 0     33 1 23  94 3 1 0 0 0
34 0 24  90 1 1 1 0 0     34 1 24 128 2 0 1 0 0
35 0 24 115 1 0 0 0 0     35 1 24 132 3 0 0 1 0
36 0 24 110 3 0 0 0 0     36 1 24 155 1 1 1 0 0
37 0 24 115 3 0 0 0 0     37 1 24 138 1 0 0 0 0
38 0 24 110 3 0 1 0 0     38 1 24 105 2 1 0 0 0
39 0 25 118 1 1 0 0 0     39 1 25 105 3 0 1 1 0
40 0 25 120 3 0 0 0 1     40 1 25  85 3 0 0 0 1
41 0 25 155 1 0 0 0 0     41 1 25 115 3 0 0 0 0
```

```
42 0 25 125 2 0 0 0 0    42 1 25  92 1 1 0 0 0
43 0 25 140 1 0 0 0 0    43 1 25  89 3 0 1 0 0
44 0 25 241 2 0 0 1 0    44 1 25 105 3 0 1 0 0
45 0 26 113 1 1 0 0 0    45 1 26 117 1 1 1 0 0
46 0 26 168 2 1 0 0 0    46 1 26  96 3 0 0 0 0
47 0 26 133 3 1 1 0 0    47 1 26 154 3 0 1 1 0
48 0 26 160 3 0 0 0 0    48 1 26 190 1 1 0 0 0
49 0 27 124 1 1 0 0 0    49 1 27 130 2 0 0 0 1
50 0 28 120 3 0 0 0 0    50 1 28 120 3 1 1 0 1
51 0 28 130 3 0 0 0 0    51 1 28  95 1 1 0 0 0
52 0 29 135 1 0 0 0 0    52 1 29 130 1 0 0 0 1
53 0 30  95 1 1 0 0 0    53 1 30 142 1 1 1 0 0
54 0 31 215 1 1 0 0 0    54 1 31 102 1 1 1 0 0
55 0 32 121 3 0 0 0 0    55 1 32 105 1 1 0 0 0
56 0 34 170 1 0 1 0 0    56 1 34 187 2 1 0 1 0
;
```

Print the first 10 observations.

```
proc print data=match_11 (obs=10);
   title 'MATCH_11 Data';
run;
```

Output 5
First 10 Observations of MATCH_11 Data Set

MATCH_11 Data

OBS	PAIR	LOW	AGE	LWT	RACE	SMOKE	PTD	HT	UI
1	1	0	14	135	1	0	0	0	0
2	1	1	14	101	3	1	1	0	0
3	2	0	15	98	2	0	0	0	0
4	2	1	15	115	3	0	0	0	1
5	3	0	16	95	3	0	0	0	0
6	3	1	16	130	3	0	0	0	0
7	4	0	17	103	3	0	0	0	0
8	4	1	17	130	3	1	1	0	1
9	5	0	17	122	1	1	0	0	0
10	5	1	17	110	1	1	0	0	0

The MATCH_NM Data Set

A large hospital-based study investigated risk factors associated with benign breast disease. Cases were women at two hospitals in New Haven, Connecticut, who had been diagnosed with benign breast disease. Controls were selected from other patients admitted to the same two hospitals. A standardized questionnaire was used to collect data from each subject in the study. The MATCH_NM data appear in Appendix 5 of Hosmer and Lemeshow (1989)* and contain these variables:

MATCHSET	stratum number
SUBJECT	observation within a matched set (1=case, 2-4=control)
AGMT	age of the subject at the interview
FNDX	diagnosis (1=case, 0=control)
CHK	regular medical checkup history (1=yes, 0=no)

* Reprinted by permission of John Wiley & Sons, Inc.

AGP1	age at first pregnancy
AGMN	age at menarche
NLV	number of stillbirths, miscarriages, and so forth
LIV	number of live births
WT	weight of the subject
AGLP	age at last menstrual period.

Program

Create the MATCH_NM data set.

```
data match_NM;
   input matchset subject agmt fndx chk agp1 agmn nlv liv wt aglp;
   datalines;
 1  1  39  1   1  23   13  0  5  118  39
 1  2  39  0   0  16   11  1  3  175  39
 1  3  39  0   0  20   12  1  3  135  39
 1  4  39  0   1  21   11  0  3  125  40
 2  1  38  1   1   .   14  .  .  118  39
 2  2  38  0   1  20   15  0  2  183  38
 2  3  38  0   1  19   11  0  5  218  38
 2  4  38  0   1  23   13  0  2  192  37
 3  1  38  1   1  22   15  2  2  125  38
 3  2  38  0   1  20   14  0  2  123  38
 3  3  38  0   1  19   13  3  2  140  37
 3  4  38  0   1  18   13  0  2  160  38
 4  1  38  1   1  24   14  2  3  150  38
 4  2  38  0   1  26   13  1  1  130  38
 4  3  38  0   0  23   14  0  4  140  38
 4  4  38  0   1  25   16  0  2  130  38
 5  1  38  1   1  21   17  0  2  150  38
 5  2  38  0   0  20   12  1  2  148  38
 5  3  38  0   1   .   13  .  .  134  39
 5  4  38  0   1  16   14  0  6  138  38
 6  1  38  1   1  24   12  1  3  116  39
 6  2  38  0   1  19   12  0  2  145  35
 6  3  38  0   1  21   10  4  3  195  35
 6  4  38  0   1  25    8  0  1  180  38
 7  1  37  1   1   .   13  .  .  137  37
 7  2  37  0   1  20   11  2  2  135  37
 7  3  37  0   1  18   10  2  3  155  37
 7  4  37  0   1  22   13  2  2  120  38
 8  1  36  1   1   .   14  .  .  126  36
 8  2  36  0   1  20   12  1  2  191  36
 8  3  36  0   0  17   10  1  3  185  37
 8  4  36  0   0  23   12  0  2  119  37
 9  1  35  1   1  23   14  0  3  129  36
 9  2  35  0   0  21   11  0  3  170  34
 9  3  36  0   1  22   14  0  4  110  36
 9  4  35  0   0  24   11  0  2  155  35
10  1  35  1   0  21   12  0  2  105  29
10  2  36  0   1  26   13  1  2  115  36
10  3  36  0   0  22   12  2  3  120  36
10  4  36  0   1  33   16  0  1  150  36
11  1  35  1   1   .   11  .  .  135  35
```

11	2	35	0	0	18	13	2	2	110	35
11	3	35	0	1	19	11	0	3	170	36
11	4	35	0	1	21	12	0	2	145	36
12	1	34	1	0	25	10	1	1	170	34
12	2	35	0	1	27	13	0	4	140	35
12	3	34	0	1	20	11	0	3	240	34
12	4	34	0	0	25	16	1	1	100	35
13	1	33	1	1	.	14	.	.	92	33
13	2	33	0	1	21	11	0	1	160	33
13	3	32	0	1	24	12	0	2	155	32
13	4	33	0	1	25	12	1	2	132	33
14	1	33	1	1	28	14	0	5	110	33
14	2	33	0	1	21	12	0	2	145	29
14	3	33	0	1	20	13	1	2	155	29
14	4	33	0	1	21	13	0	1	110	33
15	1	32	1	1	30	13	0	1	129	32
15	2	32	0	1	25	11	0	2	131	32
15	3	32	0	0	20	9	1	2	218	26
15	4	32	0	1	23	16	0	2	115	32
16	1	31	1	1	30	14	1	0	110	30
16	2	30	0	1	21	14	0	3	130	30
16	3	31	0	1	23	11	0	2	97	31
16	4	31	0	1	24	13	0	3	120	31
17	1	68	1	1	22	12	0	3	130	50
17	2	68	0	1	34	14	0	3	150	53
17	3	68	0	0	.	13	.	.	123	35
17	4	68	0	0	19	12	0	7	145	46
18	1	64	1	0	30	14	1	3	135	53
18	2	64	0	1	.	14	.	.	132	44
18	3	64	0	1	26	11	0	5	205	42
18	4	64	0	1	25	10	0	2	127	50
19	1	63	1	1	21	15	0	5	120	52
19	2	63	0	0	.	12	.	.	145	46
19	3	63	0	0	.	14	.	.	175	51
19	4	63	0	0	24	11	0	3	144	50
20	1	62	1	0	.	16	.	.	163	33
20	2	62	0	1	26	15	0	2	170	39
20	3	62	0	0	32	12	0	2	134	53
20	4	62	0	1	22	12	1	3	155	39
21	1	61	1	1	28	14	0	3	145	53
21	2	61	0	1	26	13	0	1	140	50
21	3	61	0	1	28	15	1	3	120	41
21	4	61	0	1	27	14	0	2	134	45
22	1	61	1	1	22	16	0	4	150	56
22	2	62	0	0	30	11	0	1	117	36
22	3	62	0	1	25	15	1	4	147	52
22	4	62	0	1	26	13	1	3	124	52
23	1	61	1	1	26	17	0	2	129	34
23	2	62	0	1	33	11	0	1	170	54
23	3	61	0	0	25	13	0	3	153	50
23	4	61	0	1	29	13	1	2	130	55
24	1	61	1	0	21	15	0	3	145	53
24	2	61	0	1	18	13	0	5	140	56
24	3	61	0	1	22	17	0	2	155	55
24	4	61	0	1	23	15	1	3	116	43

25	1	60	1	1	28	17	0	2	115	51
25	2	60	0	0	25	11	0	2	175	42
25	3	60	0	0	24	13	0	2	179	50
25	4	60	0	1	33	15	0	3	119	47
26	1	58	1	1	20	12	1	5	153	53
26	2	58	0	0	25	16	0	3	185	55
26	3	58	0	0	.	12	.	.	280	42
26	4	58	0	1	24	10	1	0	140	25
27	1	55	1	1	30	16	1	2	126	44
27	2	55	0	0	30	13	0	2	193	50
27	3	55	0	0	.	12	.	.	140	55
27	4	55	0	1	24	14	0	6	116	47
28	1	55	1	1	24	14	0	4	140	52
28	2	55	0	0	.	14	.	.	138	50
28	3	55	0	1	16	12	2	3	175	47
28	4	55	0	1	26	15	2	4	155	50
29	1	52	1	0	.	12	.	.	125	36
29	2	52	0	1	28	12	0	2	113	45
29	3	52	0	0	20	14	2	6	110	40
29	4	52	0	0	25	13	0	3	190	48
30	1	52	1	1	23	14	0	3	114	50
30	2	52	0	1	21	12	0	3	126	43
30	3	52	0	1	23	11	1	2	159	42
30	4	52	0	1	20	11	0	5	170	42
31	1	51	1	0	24	16	0	5	156	52
31	2	51	0	0	24	12	3	4	161	50
31	3	51	0	1	22	13	0	2	150	45
31	4	51	0	1	24	13	0	5	115	51
32	1	49	1	1	.	14	0	.	95	49
32	2	49	0	0	25	12	0	2	235	44
32	3	49	0	1	24	13	0	3	145	44
32	4	49	0	1	25	13	0	3	123	49
33	1	48	1	1	22	11	0	3	145	48
33	2	48	0	0	22	11	0	1	155	48
33	3	48	0	1	.	12	.	.	115	48
33	4	48	0	0	19	11	7	0	190	29
34	1	47	1	1	26	14	0	4	120	47
34	2	47	0	0	20	12	0	5	110	47
34	3	47	0	1	24	14	0	2	148	45
34	4	47	0	1	22	13	0	3	120	45
35	1	47	1	1	19	12	0	1	132	47
35	2	47	0	0	23	15	1	3	115	29
35	3	47	0	1	23	13	0	2	125	47
35	4	47	0	1	21	12	1	5	120	39
36	1	46	1	0	27	15	1	11	155	46
36	2	46	0	1	19	11	0	3	170	45
36	3	46	0	1	26	13	0	7	180	46
36	4	46	0	1	15	13	0	1	179	40
37	1	46	1	1	27	12	4	4	137	46
37	2	46	0	1	23	12	0	4	107	46
37	3	46	0	1	22	11	0	6	144	46
37	4	46	0	1	17	13	0	3	89	39
38	1	45	1	1	33	14	0	2	80	45
38	2	45	0	1	25	13	1	1	142	38
38	3	45	0	0	20	11	1	1	150	45

```
38   4   45   0    1  22    11   0   3  154   46
39   1   45   1    0   .    12   .   .   90   45
39   2   45   0    0  23    11   0   2  150   45
39   3   45   0    1  20    12   0   1  102   28
39   4   45   0    1  30    12   0   3  110   45
40   1   45   1    1  18    15   4   4  101   45
40   2   45   0    1  22    17   1   2  109   40
40   3   45   0    1  30    13   0   2  210   40
40   4   45   0    1  22    10   0   5  198   33
41   1   45   1    1  25    16   1   4  124   45
41   2   45   0    0  23    12   3   3  133   45
41   3   45   0    1  23    13   0   3  120   46
41   4   45   0    1  23    12   0   4  165   35
42   1   44   1    1  25    12   0   3  130   44
42   2   44   0    1  27    13   1   3  240   45
42   3   44   0    1  27    14   0   1  125   44
42   4   44   0    1   .    13   .   .  183   44
43   1   44   1    1  24    15   0   1  130   44
43   2   44   0    0  22    15   0   1  105   44
43   3   44   0    1  23    12   0   5  123   33
43   4   44   0    1  18    17   1   7  180   44
44   1   43   1    1  27    15   0   2  130   43
44   2   43   0    1  31    12   0   1  104   43
44   3   43   0    1  14    12   1   2  158   21
44   4   43   0    1  20    14   0   6  160   39
45   1   28   1    1   .    12   .   .  108   29
45   2   27   0    1  22    12   0   1  127   27
45   3   28   0    1  20    11   0   2  145   27
45   4   28   0    1  23    16   0   2  127   29
46   1   53   1    1  29    12   0   4  132   50
46   2   53   0    1  28    11   0   3  140   49
46   3   53   0    0   .    12   .   .   98   43
46   4   53   0    1  26    11   0   1  130   49
47   1   56   1    1  21    17   1   6  130   47
47   2   56   0    0  27    11   0   4  265   42
47   3   56   0    1  26    13   0   4  195   50
47   4   56   0    1  25    12   2   2  125   47
48   1   41   1    1  25    16   1   3  105   27
48   2   41   0    1  20    13   1   4  161   31
48   3   41   0    0  21    14   0   5  135   36
48   4   41   0    1  22    12   0   4  185   41
49   1   41   1    1  40    15   0   1  115   41
49   2   41   0    1  21    16   0   3  140   41
49   3   40   0    1  21    12   0   4  145   40
49   4   41   0    0  26    14   2   3  195   41
50   1   41   1    1  34    13   1   2  138   42
50   2   42   0    1   .    13   .   .  118   41
50   3   41   0    0  30    12   1   2  129   41
50   4   41   0    1  21    12   0   2  180   41
;
```

Print the first 10 observations.

```
proc print data=match_NM (obs=10);
   title 'MATCH_NM Data';
run;
```

Output 6
First 10 Observations of MATCH_NM Data Set

MATCH_NM Data

OBS	MATCHSET	SUBJECT	AGMT	FNDX	CHK	AGP1	AGMN	NLV	LIV	WT	AGLP
1	1	1	39	1	1	23	13	0	5	118	39
2	1	2	39	0	0	16	11	1	3	175	39
3	1	3	39	0	0	20	12	1	3	135	39
4	1	4	39	0	1	21	11	0	3	125	40
5	2	1	38	1	1	.	14	.	.	118	39
6	2	2	38	0	1	20	15	0	2	183	38
7	2	3	38	0	1	19	11	0	5	218	38
8	2	4	38	0	1	23	13	0	2	192	37
9	3	1	38	1	1	22	15	2	2	125	38
10	3	2	38	0	1	20	14	0	2	123	38

The MORTAL Data Set

A study of German women, many of whom were pregnant for the first time or had complications with previous pregnancies, measured their rates of perinatal mortality. The women's age, smoking habits, and the gestation period of their pregnancies were recorded. McCullagh and Nelder (1989) list a table of these data, which first appear in Wermuth (1976).

The data set contains the following variables:

DEATHS — number of deaths

TBIRTHS — total number of births

CIGS — number of cigarettes smoked per day (1=5 or less, 2=more than 5)

AGE — mother's age (1=younger than 30, 2=30 or older)

GESTPD — gestation period (1=197-260 days, 2=261 or more days).

Program

Create the MORTAL data set.

```
data mortal;
   input deaths tbirths cigs age gestpd;
   datalines;
50 365  1 1 1
9  49   2 1 1
41 188  1 2 1
4  15   2 2 1
24 4036 1 1 2
6  465  2 1 2
14 1508 1 2 2
1  125  2 2 2
;
```

Print the full data set.

```
proc print data=mortal;
   title 'Perinatal Mortality Data';
run;
```

Output 7
MORTAL Data Set

Perinatal Mortality Data

OBS	DEATHS	TBIRTHS	CIGS	AGE	GESTPD
1	50	365	1	1	1
2	9	49	2	1	1
3	41	188	1	2	1
4	4	15	2	2	1
5	24	4036	1	1	2
6	6	465	2	1	2
7	14	1508	1	2	2
8	1	125	2	2	2

The PAIRS Data Set

The paired comparison data set consists of hypothetical data for a paired comparison study of eight different types of fruit juice. Each type of juice was tasted with one other type of juice, and the taster expressed a preference for one type. No ties were allowed. A total of 60 people took part in the study. Each person tasted 28 $(8 \times 7 / 2 = 28)$ pairs of juice. The juices are coded with the letters A-H and the preferences for each juice are listed in a square matrix, which can be input as a SAS data set. Each number in the matrix represents the number of times the juice for that row was preferred to the juice for the column. For example, the value 17 in the second column of the first row indicates that 17 people preferred juice A to juice B. Correspondingly, 43 people must have preferred juice B to juice A. The value of 43 in the first column of the second row confirms this.

Program

Create the PAIRS data set.

```
data pairs;
   input a b c d e f g h;
   datalines;
 .  17 39 44  7 40 18 23
 43 .  25 39 13 17 30 35
 21 35 .  51 11 14 48 26
 16 21  9  .  3 21 18 11
 53 47 49 57  . 39 31 58
 20 43 46 39 21  . 40 22
 42 30 12 42 29 20  . 17
 37 25 34 49  2 38 43  .
;
```

Print the full data set.

```
proc print data=pairs;
   title 'Paired Comparison Data';
run;
```

Output 8
PAIRS Data Set

```
                    Paired Comparison Data

   OBS     A     B     C     D     E     F     G     H

    1      .    17    39    44     7    40    18    23
    2     43     .    25    39    13    17    30    35
    3     21    35     .    51    11    14    48    26
    4     16    21     9     .     3    21    18    11
    5     53    47    49    57     .    39    31    58
    6     20    43    46    39    21     .    40    22
    7     42    30    12    42    29    20     .    17
    8     37    25    34    49     2    38    43     .
```

The PROSTATE Data Set

The treatment for patients with prostate cancer depends on whether or not the cancer has spread to the surrounding lymph nodes. A surgical procedure (laparectomy) into the abdominal cavity can determine the extent of this nodal involvement. Certain variables may indicate nodal involvement without the need for surgery. A study was performed on 53 patients with prostate cancer to collect data on several variables considered predictive of nodal involvement. In addition, each patient had a laparectomy to determine if the cancer had spread to the lymph nodes. The data are listed in Collett (1991)* and contain the following variables:

CASE	an identification variable
AGE	age (in years) of patient at time of diagnosis
ACID	level of serum acid phosphatase in King-Armstrong units
XRAY	X-ray examination results (0=neg, 1=pos)
SIZE	size of the tumor as determined by a rectal examination (0=small, 1=large)
GRADE	summary of the pathological grade of the tumor as determined from a biopsy (0=less serious, 1=more serious)
NODALINV	laparectomy results (0=lymph nodes not involved, 1=lymph nodes involved).

Program

Create the PROSTATE data set. Create the LACD variable by taking the log of the ACID variable. Use the LOG function. This gives you better discrimination between closely spaced values.

```
data prostate;
   input case age acid xray size grade nodalinv @@;
   lacd=log(acid);
   datalines;
1  66  .48 0 0 0 0  2  68  .56 0 0 0 0  3  66  .50 0 0 0 0
4  56  .52 0 0 0 0  5  58  .50 0 0 0 0  6  60  .49 0 0 0 0
7  65  .46 1 0 0 0  8  60  .62 1 0 0 0  9  50  .56 0 0 1 1
10 49  .55 1 0 0 0  11 61  .62 0 0 0 0  12 58  .71 0 0 0 0
13 51  .65 0 0 0 0  14 67  .67 1 0 1 1  15 67  .47 0 0 1 0
16 51  .49 0 0 0 0  17 56  .50 0 0 1 0  18 60  .78 0 0 0 0
19 52  .83 0 0 0 0  20 56  .98 0 0 0 0  21 67  .52 0 0 0 0
22 63  .75 0 0 0 0  23 59  .99 0 0 1 1  24 64 1.87 0 0 0 0
```

* Reprinted by permission of the publisher.

```
25 61 1.36 1 0 0 1  26 56  .82 0 0 0 1  27 64  .40 0 1 1 0
28 61  .50 0 1 0 0  29 64  .50 0 1 1 0  30 63  .40 0 1 0 0
31 52  .55 0 1 1 0  32 66  .59 0 1 1 0  33 58  .48 1 1 0 1
34 57  .51 1 1 1 1  35 65  .49 0 1 0 1  36 65  .48 0 1 1 0
37 59  .63 1 1 1 0  38 61 1.02 0 1 0 0  39 53  .76 0 1 0 0
40 67  .95 0 1 0 0  41 53  .66 0 1 1 0  42 65  .84 1 1 1 1
43 50  .81 1 1 1 1  44 60  .76 1 1 1 1  45 45  .70 0 1 1 1
46 56  .78 1 1 1 1  47 46  .70 0 1 0 1  48 67  .67 0 1 0 1
49 63  .82 0 1 0 1  50 57  .67 0 1 1 1  51 51  .72 1 1 0 1
52 64  .89 1 1 0 1  53 68 1.26 1 1 1 1
;
```

Print the first five observations.

```
proc print data=prostate(obs=5);
   title 'Prostate Data';
run;
```

Output 9
First Five Observations of PROSTATE Data Set

Prostate Data

OBS	CASE	AGE	ACID	XRAY	SIZE	GRADE	NODALINV	LACD
1	1	66	0.48	0	0	0	0	-0.73397
2	2	68	0.56	0	0	0	0	-0.57982
3	3	66	0.50	0	0	0	0	-0.69315
4	4	56	0.52	0	0	0	0	-0.65393
5	5	58	0.50	0	0	0	0	-0.69315

References

- Collett, D.R. (1991), *Modeling Binary Data*, London: Chapman and Hall.
- Friendly, M. (1991), *SAS® System for Statistical Graphics, First Edition*, Cary, NC: SAS Institute Inc.
- Hosmer, D.W. and Lemeshow, S. (1989), *Applied Logistic Regression*, New York: John Wiley & Sons, Inc.
- McCullagh, P. and Nelder, J.A. (1989), *Generalized Linear Models, Second Edition*, London: Chapman and Hall.
- Reaven, G.M. and Miller, R.G. (1979), "An Attempt to Define the Nature of Chemical Diabetes Using a Multidimensional Analysis," *Diabetologia*, 16, 17-24.
- Wermuth, N. (1976), "Exploratory Analyses of Multidimensional Contingency Tables," *Proceedings of the Ninth International Biometrics Conference*, Volume I, 279-295.

EXAMPLE 1

Fitting a Binary Logistic Regression Model

Featured Tools:
LOGISTIC procedure:

- □ DESCENDING option
- □ MODEL statement

Logistic regression analysis models the relationship between a binary or ordinal response variable and one or more explanatory variables. The logistic regression model uses the explanatory variables to predict the probability that the response variable takes on a given value. The response variable can take on one of two binary values or one of a (usually small) number of ordinal values.

Response variables that can take on a large number of different values — such as variables measured on an interval or ratio scale — are usually modeled with simple linear regression, rather than with logistic regression.

For a binary response variable y, the linear logistic regression model has the form

$$\mathrm{logit}\left(p_i\right) = \log\left(p_i / \left(1 - p_i\right)\right) = \alpha + \beta'\mathbf{x}_i$$

where

$p_i = \mathrm{Prob}(y_i = y_1 \mid \mathbf{x}_i)$	is the response probability to be modeled, and y_1 is the first ordered level of y.
α	is the intercept parameter.
β	is the vector of slope parameters.
$\mathbf{x}_i$	is the vector of explanatory variables.

This logistic regression equation models the logit transformation of the ith individual's event probability, p_i, as a linear function of the explanatory variables in the vector, $\mathbf{x}_i$. A more general class of models share the feature that a function $g = g(\mu)$ of the mean of the response variable is assumed to be linearly related to the explanatory variables. The function g is called the *link function.* Other common link functions are the normit function (used in probit analysis) and the complementary log-log function. The logit function has the advantage of being more easily interpreted. Also, it can be applied to data that are collected prospectively or retrospectively.

Program

Create the ESR data set. The complete ESR data set is in the Introduction.

```
data esr;
   input id fibrin globulin response @@;
   datalines;
1  2.52 38 0   2  2.56 31 0   3  2.19 33 0   4  2.18 31 0
more data lines
;
```

Fit the logistic regression model.
❶ Assign proper response level ordering with the DESCENDING option.

```
proc logistic data=esr descending;
   model response=fibrin globulin;
   title 'ESR Data';
run;
```

Output

Output 1.1
Output Generated by PROC LOGISTIC

```
                              ESR Data

                       The LOGISTIC Procedure

      Data Set: WORK.ESR
      Response Variable: RESPONSE
      Response Levels: 2
      Number of Observations: 32
      Link Function: Logit

                     ❶    Response Profile

                      Ordered
                        Value  RESPONSE     Count

                            1         1         6
                            2         0        26

                       The LOGISTIC Procedure

          ❷ Testing Global Null Hypothesis: BETA=0

                                 Intercept
                   Intercept        and
    Criterion        Only       Covariates    Chi-Square for Covariates

    AIC             32.885        28.971           .
    SC              34.351        33.368           .
    -2 LOG L        30.885        22.971         7.914 with 2 DF (p=0.0191)
    Score              .             .           8.207 with 2 DF (p=0.0165)

                       The LOGISTIC Procedure

                 Analysis of Maximum Likelihood Estimates
                  ❸        ❹        ❺         ❻          ❼            ❽
              Parameter Standard    Wald      Pr >    Standardized    Odds
Variable DF   Estimate   Error  Chi-Square Chi-Square   Estimate      Ratio

INTERCPT 1    -12.7921   5.7964     4.8704     0.0273          .          .
FIBRIN   1      1.9104   0.9710     3.8708     0.0491   0.670987      6.756
GLOBULIN 1      0.1558   0.1195     1.6982     0.1925   0.393641      1.169

 ❾  Association of Predicted Probabilities and Observed Responses

               Concordant = 80.1%          Somers' D = 0.609
               Discordant = 19.2%          Gamma     = 0.613
               Tied       =  0.6%          Tau-a     = 0.192
               (156 pairs)                 c         = 0.804
```

Explanation

The LOGISTIC procedure prints tables and statistics to help you analyze and evaluate the estimated linear logistic regression model.

❶ For each response level, the `Response Profile` table gives the ordered value, the value of the response variable, and the count or frequency. The event observations are ordered value 1 and the nonevent observations are ordered value 2.

❷ The `Testing Global Null Hypothesis: BETA=0` table lists two criteria (AIC and SC) that are useful for comparing models, and two criteria (-2 LOG L and Score) that test the null hypothesis that all regression coefficients are zero. Except for the score statistic, all of the criteria are based on the likelihood for fitting a model with intercepts only or for fitting a model with intercepts and explanatory variables. (The explanatory variables are also called independent variables or covariates.)

- `AIC` is the Akaike Information Criterion, which is an adjustment to the -2 LOG L score based on the number of explanatory variables in the model and the number of observations used. For a given set of data, the AIC is a goodness-of-fit measure that you can use to compare one model to another, with lower values indicating a more desirable model.
- `SC` is the Schwarz Criterion, which is another way of adjusting the -2 LOG L score based on the number of explanatory variables in the model and the number of observations used. For a given set of data, the SC is also a goodness-of-fit measure that you can use to compare one model to another, with lower values indicating a more desirable model.
- `-2 LOG L` is the -2 Log Likelihood statistic, which has a chi-square distribution under the null hypothesis that all regression coefficients of the model are zero. The procedure prints this chi-square value in the third column of the table, along with the degrees of freedom, and a *p*-value for this statistic. A significant *p*-value (for example, a *p*-value less than .05) provides evidence that at least one of the regression coefficients for an explanatory variable is nonzero.
- `Score` is a score statistic, which has an asymptotic chi-square distribution under the null hypothesis. The procedure prints the chi-square value, degrees of freedom, and a *p*-value for this statistic.

❸ `Parameter Estimate` gives the estimated coefficients of the fitted logistic regression model. Here, the logistic regression equation is

$$\text{logit}(\hat{p}) = -12.79 + 1.91 \times \text{FIBRIN} + .16 \times \text{GLOBULIN}$$

The hat over the p indicates that it is an estimated probability value for the response variable. The estimated slope coefficients indicate that you expect the logit transformation of the event probability to increase by 1.91 for each unit increase in FIBRIN and you expect it to increase by .16 for each unit increase in GLOBULIN.

❹ `Standard Error` gives the standard error of the parameter estimates.

❺ **`Wald Chi-Square`** is computed as the square of the value obtained by dividing the parameter estimate by its standard error. That is, the Wald chi-square is computed from the following formula:

$$\left(\frac{\hat{\beta}}{\text{s.e.}(\hat{\beta})}\right)^2$$

❻ **`Pr > Chi-Square`** is the *p*-value for the Wald chi-square statistic with respect to a chi-square distribution with one degree of freedom. Note that the *p*-value for GLOBULIN is .1925, which is nonsignificant at the .05 or .10 level.

❼ **`Standardized Estimate`** is computed for all slope parameters with the following formula:

$$\frac{\hat{\beta}_x}{\frac{\sqrt{\pi^2 / 3}}{\sigma_x}}$$

That is, divide the slope parameter estimate by the ratio of the standard deviation of the underlying distribution to the sample standard deviation of the explanatory variable. Note that $\sqrt{\pi^2 / 3}$ is the standard deviation of the underlying logistic distribution. Specify the SIMPLE option in the PROC LOGISTIC statement to print simple statistics for each explanatory variable. From a listing of the simple statistics, you find that the sample standard deviation of the FIBRIN variable is .637.

The standardized estimates of the intercept parameters are set to missing.

❽ **`Odds Ratio`** is computed by exponentiating the parameter estimate for each explanatory variable. In this example, the odds of the event increase by a factor of 6.756 for each unit increase in the value of FIBRIN, and the odds increase by a factor of 1.169 for each unit increase in the value of GLOBULIN.

❾ The **`Association of Predicted Probabilities and Observed Responses`** table lists several measures of association to help you assess the quality of the logistic model.

The procedure gives the percentage of concordant, discordant, and tied observations, and the number of observation pairs upon which the percentages are based. For all pairs of observations with different values of the response variable, a pair is *concordant* if the observation with the larger ordered value of the response has a lower predicted event probability than does the observation with the smaller ordered value of the response. Recall that the event is ordered value 1 of the response and the nonevent is ordered value 2 of the response. A pair is *discordant* if the observation with the larger ordered value of the response has a higher predicted event probability than does the observation with the smaller ordered value of the response. If a pair is neither concordant or discordant, it is a *tie*. In this example, there are 26 observations with a response value of 0, and 6 observations with a response value of 1, which creates a total of $26 \times 6 = 156$ pairs of observations with different response values. 80.1 percent of these pairs are concordant.

The four rank correlation indexes in the table are computed from the numbers of concordant and discordant pairs of observations. In a relative sense, a model with higher values for these indexes has better predictive ability than a model with lower values for these indexes. The indexes use the following formulas:

Somer's D = $(nc - nd) / t$

Gamma = $(nc - nd) / (nc + nd)$

Tau-a = $(nc - nd) / .5N(N - 1)$

c = $(nc + .5(t - nc - nd)) / t$

where

N is the total number of observations in the input data set.

t is the total number of pairs with different response values.

nc is the number of concordant pairs.

nd is the number of discordant pairs.

Note two items about these measures of association:

- Somer's D is equal to 2(c − .5).
- c is equal to the area under a receiver operating characteristic curve (Bamber 1975; Hanley and McNeil 1982).

A Closer Look

Assign Proper Response Level Ordering

By default, PROC LOGISTIC models the probability of the response that corresponds to the lower ordered value. For example, if your response variable has the value 0 for nonevents and the value 1 for events, then PROC LOGISTIC models the probability of the nonevent. This can be confusing when you try to interpret the results of the logistic regression analysis. In the case where PROC LOGISTIC models the probability of the nonevent, the parameter estimates for the explanatory variables indicate how increasing or decreasing values of those variables affect the probability of the nonevent. In most cases, however, you want to know how the probability of the event of interest, rather than the nonevent, is affected by increasing or decreasing values of the explanatory variables.

CAUTION!

It is crucial that you examine the Response Profile table in the PROC LOGISTIC output to verify that the response level representing the event of interest has ordered value 1. ■

There are several ways to reverse the response level ordering:

- Use the DESCENDING option in the PROC LOGISTIC statement to reverse the default ordering of the response variable, as shown in the "Program" section of this example. This is the simplest method of reversing the response level ordering. The DESCENDING option is available in Release 6.07 TS301 and later releases.
- Assign a format to the response variable such that the first formatted value (when the formatted values are put in sorted order) corresponds to the event of interest. For this example, the response value of 0 is assigned the

formatted value '**No ESR**', and the response value of 1 is assigned the formatted value '**ESR**':

```
proc format;
   value esrfmt 0='No ESR' 1='ESR';
run;

proc logistic data=esr;
   model response=fibrin globulin;
   format response esrfmt.;
run;
```

- Sort the input data so that at least one observation with the response value of the event of interest occurs before observations with the value of the nonevent. Then use the ORDER=DATA specification in the PROC LOGISTIC statement. The following example shows how to sort the ESR data to use this method. Recall that for the ESR data the event of interest has the larger value.

```
proc sort data=esr;
   by descending response;
run;

proc logistic data=esr order=data;
   model response=fibrin globulin;
run;
```

- Create a new variable to replace the response variable in the MODEL statement of PROC LOGISTIC such that the event of interest is represented by the smaller value of the new variable:

```
data esr2;
   set esr;
   if response=0 then resp2='No ESR';
   else resp2='ESR';
run;

proc logistic data=esr2;
   model resp2=fibrin globulin;
run;
```

- Create a new variable with constant value 1 for each observation. Use the events/trials MODEL statement syntax in PROC LOGISTIC with the original response variable as the event variable and the new variable as the trial variable. See Example 10 for more information on the events/trials syntax. Note that this method depends on the original response variable having values of 0 and 1, with 1 corresponding to the event of interest:

```
data esr3;
   set esr;
   retain n 1;
run;
```

```
proc logistic data=esr3;
   model response/n=fibrin globulin;
run;
```

References

- Bamber, D. (1975), "The Area Above the Ordinal Dominance Graph and the Area Below the Receiver Operating Characteristic Graph," *Journal of Mathematical Psychology*, 12, 387-415.
- Hanley, J.A. and McNeil, B.J. (1982), "The Meaning and Use of the Area Under a Receiver Operating Characteristic (ROC) Curve," *Radiology*, 143, 29-36.

E X A M P L E 2

Computing Confidence Limits for Regression Parameters and Odds Ratios

Featured Tools:
PROC LOGISTIC, MODEL statement:

- ALPHA= option
- PLCL option
- PLRL option
- WALDCL option
- WALDRL option

Estimated odds ratios are computed by exponentiating the parameter estimates for a logistic regression model when the following conditions are met:

- the explanatory variable does not interact with any other variable
- the explanatory variable is represented by a single term in the model
- a one-unit change in the explanatory variable is relevant.

Similarly, confidence limits for odds ratios are computed by exponentiating the confidence limits for the logistic regression parameters.

There are two available methods of computing confidence limits for logistic regression parameters: the likelihood ratio method and the Wald method. The *likelihood ratio method* is an iterative process based on the profile likelihood function. The *Wald method* is a simpler method based on the asymptotic normality of the parameter estimator. These two methods should produce approximately the same results for large samples, but may produce different results for small samples. When the parameter estimate is very large, however, these two methods may produce different results even for large sample sizes.

This example uses options in the MODEL statement of the LOGISTIC procedure to compute confidence limits for odds ratios and parameter estimates. It also shows how to use an option to adjust the confidence coefficient for the confidence limits.

Note: The PLCL, PLRL, WALDCL, and WALDRL options are available in Release 6.10 and later releases. WALDRL is equivalent to the RISKLIMITS option, which is available in earlier releases.

Program

Create the DIABETES data set. The complete DIABETES data set is in the Introduction.

```
data diabetes;
   input patient relwt glufast glutest instest sspg group;
   label relwt   = 'Relative weight'
         glufast = 'Fasting Plasma Glucose'
         glutest = 'Test Plasma Glucose'
         instest = 'Plasma Insulin during Test'
         sspg    = 'Steady State Plasma Glucose'
         group   = 'Clinical Group';
   datalines;
1  0.81  80  356 124   55  1
2  0.95  97  289 117   76  1
3  0.94 105  319 143  105  1
more data lines
;
```

Convert the ordinal response (GROUP) to a binary response (GRP). Combine the chemical diabetics and overt diabetics into one group — the event group. The normals are the nonevent group.

```
data diabet2;
   set diabetes;
   grp=(group=1);
run;
```

Compute both types of confidence limits for regression parameter estimates and odds ratios. Use options in the MODEL statement.

```
proc logistic data=diabet2;
   model grp=glutest / plcl plrl waldcl waldrl;
   title 'Diabetes Data';
run;
```

Output

Output 2.1
Partial Output Generated by PROC LOGISTIC

```
                         Diabetes Data

                    The LOGISTIC Procedure

❶ Parameter Estimates and 95% Confidence Intervals

                                      Profile Likelihood
                                      Confidence Limits
                     Parameter
      Variable       Estimate           Lower        Upper

      INTERCPT       -90.4017          -213.2     -38.6425
      GLUTEST          0.2153          0.0918       0.5073

❷ Parameter Estimates and 95% Confidence Intervals

                                              Wald
                                      Confidence Limits
                     Parameter
      Variable       Estimate           Lower        Upper

      INTERCPT       -90.4017          -173.8      -7.0490
      GLUTEST          0.2153          0.0171       0.4136

❸ Conditional Odds Ratios and 95% Confidence Intervals

                                               Profile Likelihood
                                               Confidence Limits
                                   Odds
   Variable          Unit         Ratio          Lower        Upper

   GLUTEST         1.0000         1.240          1.096        1.661
```

```
❹ Conditional Odds Ratios and 95% Confidence Intervals

                                              Wald
                                        Confidence Limits
                               Odds
Variable        Unit          Ratio       Lower       Upper

GLUTEST       1.0000          1.240       1.017       1.512
```

Explanation

❶ The PLCL option produces the first table labeled **`Parameter Estimates and 95% Confidence Intervals`** in Output 2.1. The confidence limits are labeled **`Profile Likelihood Confidence Limits`**. The construction of these confidence intervals is derived from the asymptotic chi-square distribution of the likelihood ratio test.

❷ The WALDCL option produces the second table labeled **`Parameter Estimates and 95% Confidence Intervals`**. The confidence limits are labeled **`Wald Confidence Limits`**. Wald confidence limits are computed by assuming a normal distribution for each parameter estimator. This computation method is less time consuming than the one based on the profile likelihood function because it does not involve an iterative process. However, it is considered to be less accurate, especially for small sample sizes.

When you examine the confidence intervals for the parameter estimates, you can see that the Wald confidence intervals are symmetric about the point estimate, but the profile likelihood confidence intervals are asymmetrical. This is because the upper and lower profile likelihood confidence limits are computed separately using an iterative process, and the distribution of a parameter estimate is not symmetric for small sample sizes.

❸ The PLRL option produces the first table labeled **`Conditional Odds Ratios and 95% Confidence Intervals`**. The confidence limits are labeled **`Profile Likelihood Confidence Limits`**. Profile likelihood confidence limits for odds ratios are a transformation of the confidence limits that you can produce with the PLCL option for the corresponding regression parameters.

❹ The WALDRL option produces the second table labeled **`Conditional Odds Ratios and 95% Confidence Intervals`**. The confidence limits are labeled **`Wald Confidence Limits`**. WALDRL is an alias of the RISKLIMITS option, which is available in Release 6.07 and later releases. It requests confidence intervals for the odds ratios of all explanatory variables. Computation of these confidence intervals is based on the asymptotic normality of the parameter estimators.

Variation

Adjusting the Confidence Coefficient

By default, the PLCL, PLRL, WALDCL, and WALDRL options compute 95% confidence limits. Use the ALPHA= option to adjust the confidence coefficient to the desired level. For example, specifying ALPHA = .01 causes the procedure to compute 99% confidence limits. Output 2.2 shows the new 99% confidence limits. The 99% confidence intervals are wider than the corresponding 95% confidence intervals shown in Output 2.1.

Adjust the confidence coefficient for all confidence intervals. Use the ALPHA= option in the MODEL statement.

```
proc logistic data=diabet2;
   model grp=glutest / plcl plrl waldcl waldrl alpha=.01;
run;
```

Output 2.2
Partial Output Generated by PROC LOGISTIC

```
                     Diabetes Data

                 The LOGISTIC Procedure

     Parameter Estimates and 99% Confidence Intervals

                                  Profile Likelihood
                                   Confidence Limits
                    Parameter
       Variable      Estimate        Lower       Upper

       INTERCPT      -90.4017       -268.6    -30.8411
       GLUTEST         0.2153       0.0731      0.6389

     Parameter Estimates and 99% Confidence Intervals

                                         Wald
                                   Confidence Limits
                    Parameter
       Variable      Estimate        Lower       Upper

       INTERCPT      -90.4017       -199.9     19.1424
       GLUTEST         0.2153      -0.0452      0.4759

   Conditional Odds Ratios and 99% Confidence Intervals

                                        Profile Likelihood
                                         Confidence Limits
                                 Odds
  Variable         Unit         Ratio        Lower        Upper

  GLUTEST        1.0000         1.240        1.076        1.894

   Conditional Odds Ratios and 99% Confidence Intervals

                                               Wald
                                         Confidence Limits
                                 Odds
  Variable         Unit         Ratio        Lower        Upper

  GLUTEST        1.0000         1.240        0.956        1.609
```

EXAMPLE 3

Computing Customized Odds Ratios

Featured Tools:

- PROC LOGISTIC:
  - NOPRINT option
  - OUTEST= option
  - UNITS statement
- DATA step processing

The slope coefficient β_j associated with an explanatory variable X_j represents the change in log odds for an increase of one unit in X_j. The odds ratio (e^{β}) is the ratio of odds for a one-unit change in X_j. Recall that the odds of an event are equal to $p / (1 - p)$, and the log odds of an event are equal to $\text{logit}(p) = \log(p / (1 - p))$, where p is the probability of the event.

The change in log odds, and the corresponding change in the odds ratio, for some amount other than one unit is often of greater interest. For example, a change of one pound in body weight may be too small to be considered important while a change of 10 pounds may be more meaningful. For a change of c units in X_j, the customized odds ratio is estimated by $\exp[c\hat{\beta}_j]$.

The corresponding lower and upper confidence limits for a customized odds ratio are $\exp[c\text{L}_j]$ and $\exp[c\text{U}_j]$, respectively, for $(c > 0)$, or $\exp[c\text{U}_j]$ and $\exp[c\text{L}_j]$, respectively, for $(c < 0)$, where (L_j,U_j) can be either the likelihood ratio-based confidence interval or the Wald confidence interval for β_j.

This example uses the UNITS statement (available in Release 6.10 and later releases) to compute customized odds ratios in the LOGISTIC procedure. It also uses the DATA step to compute customized odds ratios from the parameter estimates that PROC LOGISTIC computes.

Program

Create the DIABETES data set. The complete DIABETES data set is in the Introduction.

```
data diabetes;
   input patient relwt glufast glutest instest sspg group;
   label relwt   = 'Relative weight'
         glufast = 'Fasting Plasma Glucose'
         glutest = 'Test Plasma Glucose'
         instest = 'Plasma Insulin during Test'
         sspg    = 'Steady State Plasma Glucose'
         group   = 'Clinical Group';
   datalines;
1  0.81  80  356 124   55  1
2  0.95  97  289 117   76  1
3  0.94 105  319 143  105  1
more data lines
;
```

Convert the ordinal response (GROUP) to a binary response (GRP). Combine the chemical diabetics and overt diabetics into one group — the event group. The normals are the nonevent group.

```
data diabet2;
   set diabetes;
   grp=(group=1);
run;
```

Compute customized odds ratios. Use the UNITS statement. The PLRL and WALDRL options compute confidence limits for this odds ratio.

```
proc logistic data=diabet2;
   model grp=glutest / plrl waldrl;
   units glutest=10 -10;
   title 'Diabetes Data';
run;
```

Output

Output 3.1
Partial Output Generated by PROC LOGISTIC

```
                         Diabetes Data

                    The LOGISTIC Procedure

           ❶ Analysis of Maximum Likelihood Estimates

          Parameter Standard    Wald       Pr >   Standardized    Odds
Variable DF  Estimate    Error Chi-Square Chi-Square   Estimate     Ratio
                                                                    ❷
INTERCPT 1   -90.4017  42.5277    4.5187     0.0335          .         .
GLUTEST  1     0.2153   0.1011    4.5317     0.0333   37.626552     1.240
```

```
          Conditional Odds Ratios and 95% Confidence Intervals

                                         ❹ Profile Likelihood
                                            Confidence Limits
                                 ❸ Odds
      Variable         Unit        Ratio        Lower       Upper

      GLUTEST        10.0000       8.613        2.505     159.644
      GLUTEST       -10.0000       0.116        0.006       0.399
```

```
                                     ❺         Wald
                                          Confidence Limits
                                   Odds
Variable          Unit             Ratio      Lower       Upper

GLUTEST         10.0000            8.613      1.186      62.534
GLUTEST        -10.0000            0.116      0.016       0.843
```

Explanation

Output 3.1 shows the maximum likelihood estimates ❶ for the logistic regression model of the diabetes data. The estimated odds ratio is 1.24 ❷, which represents the multiplicative change in odds for a one-unit increase in the value of GLUTEST. Considering the values of the GLUTEST variable, however, it is of more interest to compute the increase in risk for a larger change in GLUTEST. In this example, you use the UNITS statement to compute new customized odds ratios for an increase of 10 units and a decrease of 10 units in GLUTEST. The customized odds ratios have a value of 8.613 for an increase of 10 units in GLUTEST, and a value of 0.116 for a decrease of 10 units in GLUTEST ❸. That is, an increase of 10 units in the value of GLUTEST multiplies the odds of being a chemical or overt diabetic 8.613 times, and a decrease of 10 units in the value of GLUTEST multiplies the odds of being a chemical or overt diabetic 0.116 times. Note that 0.116 is equal to 1/8.613. The PLRL and WALDRL options compute the confidence limits that use the likelihood-ratio method ❹ and the Wald method ❺, respectively. See Example 2 for more information on these options and on computing confidence limits.

Variation

Computing Customized Odds Ratios in the DATA Step

You can compute customized odds ratios easily in a DATA step, as shown in the following example. You create a standard odds ratio (OR) that represents a change of one unit in the GLUTEST variable. You also create customized odds ratios (ORCUST1 and ORCUST2) that represent an increase of 10 units and a decrease of 10 units, respectively, in the value of GLUTEST.

Create an output data set that contains the final parameter estimates. Use the OUTEST= option. The NOPRINT option suppresses all printed output.

```
proc logistic data=diabet2
              outest=stats1
              noprint;
   model grp=glutest;
run;
```

Print the OUTEST= data set.

```
proc print data=stats1;
run;
```

Output 3.2
The OUTEST= Data Set

Diabetes Data

OBS	_LINK_	_TYPE_	_NAME_	INTERCEP	GLUTEST	_LNLIKE_
1	LOGIT	PARMS	ESTIMATE	-90.4017	0.21532	-5.55189

Compute odds ratios. Use the EXP function.

```
data stats2(keep=glutest or orcust1 orcust2);
   set stats1;
   or=exp(glutest);
   orcust1=exp(10*glutest);
   orcust2=exp(-10*glutest);
run;
```

Print the data set that contains the customized odds ratios.

```
proc print data=stats2;
run;
```

Output 3.3
Customized Odds Ratios Created in a DATA Step

Diabetes Data

OBS	GLUTEST	OR	ORCUST1	ORCUST2
1	0.21532	1.24026	8.61269	0.11611

EXAMPLE 4

Computing Predicted Probabilities and Classifying Observations

Featured Tools:

- PROC LOGISTIC, OUTPUT statement:

  OUT= option

  PREDICTED= option
- FREQ procedure

You can obtain predicted values from a logistic regression model, just as you can obtain predictions from a classical linear regression model. The predicted probability from a binary logistic regression model is the estimated probability that an observation is an event. For example, in an epidemiological study, this might represent the probability that a given individual has the disease being studied.

The predicted probabilities, $\hat{p}$, can be computed from the following formula:

$$\hat{p} = \frac{1}{1 + \exp\left(\hat{\alpha} - \hat{\beta}'\mathbf{x}\right)}$$

where

$\hat{\alpha}$ is the intercept parameter estimate.

$\hat{\beta}$ is the vector of slope parameter estimates.

$\mathbf{x}$ is the vector of explanatory variables.

After computing the predicted probabilities, you may be interested in how well your model predicts the actual incidences of the event. The best methods for doing this relate the vector of actual predicted probabilities to the vector of binary responses. The log likelihood ratio chi-square is one measure, but its value is proportional to the sample size. Unitless indexes of predictive accuracy include the Brier score, which is expressed by the following formula: $\text{avg}\left[\left(\hat{p} - y\right)^2\right]$ (Brier 1950), and the area under the receiver operating characteristic (ROC) curve (Bamber 1975; Hanley and McNeil 1982). The Brier score is a strictly proper scoring rule, which means that it is minimized for predicted probabilities that are equal to the true probabilities. Example 10 provides an example and more discussion of ROC curves. A statistically insensitive method to measure predictive accuracy, but one that is easy to understand, is based on categorizing predictions. You can use a probability cutpoint, such as .50, to classify an observation as a predicted event or a predicted nonevent. Then, you can compare the predicted events to the actual observed responses and get a measure of the predictive accuracy of your model. Example 5 shows how to classify observations over a large range of probability cutpoints.

When you use the same data to test the predictive accuracy of your model that you use to fit the model, it biases the results. Example 4 shows one way to avoid this bias: to use a new set of observations to test the predictive accuracy of your model. Example 5 shows another way to avoid this bias, which is to fit a model that omits each observation one at a time and then classify each observation as an event or nonevent based on the model that omits the observation being classified. The method shown in Example 5 is called *jackknifing.*

This example uses options in the OUTPUT statement of the LOGISTIC procedure to compute predicted probabilities from a logistic regression model and write them to an output data set. It uses the FREQ procedure to produce a classification table for a single probability cutpoint. This example also shows how to compute a Brier score for testing the predictive accuracy of a model.

Program

Create the PROSTATE data set. The complete PROSTATE data set is in the Introduction.

```
data prostate;
   input case age acid xray size grade nodalinv @@;
   lacd=log(acid);
   datalines;
1 66  .48 0 0 0 0   2 68  .56 0 0 0 0   3 66  .50 0 0 0 0
more data lines
;
```

Create an output data set that contains the predicted probabilities for each observation in the input data set. Use the OUTPUT statement with the PREDICTED= option.

```
proc logistic data=prostate descending noprint;
   model nodalinv=lacd xray size;
   output out=probs predicted=phat;
run;
```

Print the output data set.

```
proc print data=probs;
   title 'Prostate Data';
run;
```

Output

Output: PROC PRINT

Output 4.1
Listing of Output Data Set from PROC LOGISTIC

Prostate Data

OBS	CASE	AGE	ACID	XRAY	SIZE	GRADE	NODALINV	LACD	_LEVEL_	PHAT
1	1	66	0.48	0	0	0	0	-0.73397	1	0.05306
2	2	68	0.56	0	0	0	0	-0.57982	1	0.07389
3	3	66	0.50	0	0	0	0	-0.69315	1	0.05796
4	4	56	0.52	0	0	0	0	-0.65393	1	0.06307
5	5	58	0.50	0	0	0	0	-0.69315	1	0.05796
6	6	60	0.49	0	0	0	0	-0.71335	1	0.05549
7	7	65	0.46	1	0	0	0	-0.77653	1	0.28407
8	8	60	0.62	1	0	0	0	-0.47804	1	0.44025
9	9	50	0.56	0	0	1	1	-0.57982	1	0.07389
10	10	49	0.55	1	0	0	0	-0.59784	1	0.37408
11	11	61	0.62	0	0	0	0	-0.47804	1	0.09153
12	12	58	0.71	0	0	0	0	-0.34249	1	0.12085
13	13	51	0.65	0	0	0	0	-0.43078	1	0.10094
14	14	67	0.67	1	0	1	1	-0.40048	1	0.48442
15	15	67	0.47	0	0	1	0	-0.75502	1	0.05069
16	16	51	0.49	0	0	0	0	-0.71335	1	0.05549
17	17	56	0.50	0	0	1	0	-0.69315	1	0.05796
18	18	60	0.78	0	0	0	0	-0.24846	1	0.14568
19	19	52	0.83	0	0	0	0	-0.18633	1	0.16431
20	20	56	0.98	0	0	0	0	-0.02020	1	0.22345
21	21	67	0.52	0	0	0	0	-0.65393	1	0.06307
22	22	63	0.75	0	0	0	0	-0.28768	1	0.13484
23	23	59	0.99	0	0	1	1	-0.01005	1	0.22751
24	24	64	1.87	0	0	0	0	0.62594	1	0.55859
25	25	61	1.36	1	0	0	1	0.30748	1	0.82642
26	26	56	0.82	0	0	0	1	-0.19845	1	0.16053
27	27	64	0.40	0	1	1	0	-0.91629	1	0.17712
28	28	61	0.50	0	1	0	0	-0.69315	1	0.26416
29	29	64	0.50	0	1	1	0	-0.69315	1	0.26416
30	30	63	0.40	0	1	0	0	-0.91629	1	0.17712
31	31	52	0.55	0	1	1	0	-0.59784	1	0.30874
32	32	66	0.59	0	1	1	0	-0.52763	1	0.34410
33	33	58	0.48	1	1	0	1	-0.73397	1	0.71849
34	34	57	0.51	1	1	1	1	-0.67334	1	0.74572
35	35	65	0.49	0	1	0	1	-0.71335	1	0.25525

```
36  36   65 0.48   0   1   1    0   -0.73397   1   0.24637
37  37   59 0.63   1   1   1    0   -0.46204   1   0.82640
38  38   61 1.02   0   1   0    0    0.01980   1   0.64789
39  39   53 0.76   0   1   0    0   -0.27444   1   0.48383
40  40   67 0.95   0   1   0    0   -0.05129   1   0.60988
41  41   53 0.66   0   1   1    0   -0.41552   1   0.40418
42  42   65 0.84   1   1   1    1   -0.17435   1   0.90201
43  43   50 0.81   1   1   1    1   -0.21072   1   0.89439
44  44   60 0.76   1   1   1    1   -0.27444   1   0.87978
45  45   45 0.70   0   1   1    1   -0.35667   1   0.43704
46  46   56 0.78   1   1   1    1   -0.24846   1   0.88593
47  47   46 0.70   0   1   0    1   -0.35667   1   0.43704
48  48   67 0.67   0   1   0    1   -0.40048   1   0.41251
49  49   63 0.82   0   1   0    1   -0.19845   1   0.52735
50  50   57 0.67   0   1   1    1   -0.40048   1   0.41251
51  51   51 0.72   1   1   0    1   -0.32850   1   0.86604
52  52   64 0.89   1   1   0    1   -0.11653   1   0.91312
53  53   68 1.26   1   1   1    1    0.23111   1   0.95888
```

Explanation

Output 4.1 lists the output data set from PROC LOGISTIC. Note that it contains all of the variables from the original input data set, not just the variables listed in the MODEL statement. In addition, the output data set contains two new variables, _LEVEL_ and PHAT. For binary logistic regression, the value of _LEVEL_ is equal to the value of the response variable being modeled. That is, the value of _LEVEL_ is equal to the ordered value 1 from the `Response Profile` table in the PROC LOGISTIC output. The predicted probabilities in the PHAT variable represent the probability that each observation is in the response level indicated by the _LEVEL_ variable. In this example, the predicted probabilities in the PHAT variable represent the probability that the response variable NODALINV has the value 1.

Creating a Classification Table with PROC FREQ

To compare the predicted events with the actual observed events, you create a table. For binary logistic regression, you create a 2 x 2 table of predicted by observed responses. First, however, you need to add a new variable to the output data set to represent the predicted binary responses. This example uses .50 as the probability cutpoint. When the value of the estimated probability is greater than or equal to .50, the observation is classified as a predicted event. If the estimated probability is less than .50, then the observation is classified as a predicted nonevent.

Program (continued)

Revise the output data set from PROC LOGISTIC. Create the variable PREDICTS, which has the value 1 when the event probability is greater than or equal to .50 and has the value 0 otherwise.

```
data probs1;
   set probs;
   predicts=(phat>=.5);
run;
```

Create the classification table from the revised output data set from PROC LOGISTIC.

```
proc freq data=probs1;
   tables nodalinv*predicts / norow nocol nopercent;
run;
```

Output: Classification Table

Output 4.2.
Classification Table for Prostate Data

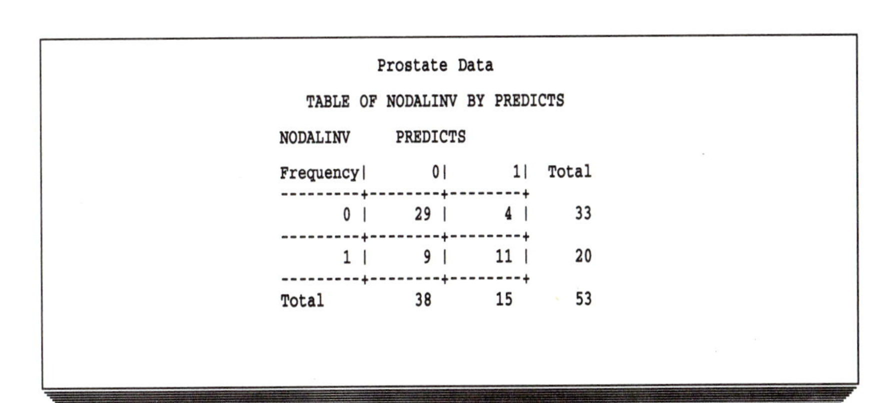

```
                Prostate Data

         TABLE OF NODALINV BY PREDICTS

      NODALINV     PREDICTS

      Frequency|       0|       1|  Total
      ---------+--------+--------+
             0 |     29 |      4 |     33
      ---------+--------+--------+
             1 |      9 |     11 |     20
      ---------+--------+--------+
      Total          38       15       53
```

Explanation

From Output 4.2, you can see that (29 + 11) = 40, or 75 percent, of the 53 observations in the prostate data are correctly classified by the logistic regression model. Of the 20 observed events, only 11, or 55 percent, are correctly classified as predicted events. Nine of these observations are incorrectly classified as predicted nonevents. These nine observations are called false negatives. Only four of the observed nonevents are incorrectly classified as predicted events. These four observations are called false positives.

Note: These results are biased because you use the same set of data to create the classification table that you use to fit the model. Example 5 shows how to use a jackknife technique to create an unbiased classification table.

One way to lower the number of false negatives is to lower the probability cutpoint for classifying an observation as a predicted event. Note that lowering the probability cutpoint increases the number of false positives as it lowers the number of false negatives. For example, if the probability cutpoint is set at .25 instead of .50, the number of false negatives decreases from 9 to 3, but the number of false positives increases from 4 to 13, as shown in the following program and output.

Program (continued)

Set the probability cutpoint to .25.

```
data probs2;
   set probs;
   predicts=(phat>=.25);
run;
```

Create the classification table.

```
proc freq data=probs2;
   tables nodalinv*predicts / norow nocol nopercent;
run;
```

Output: Revised Classification Table

Output 4.3
Revised Classification Table for Prostate Data

```
                 Prostate Data

          TABLE OF NODALINV BY PREDICTS

      NODALINV     PREDICTS

      Frequency|       0|       1|  Total
      ---------+--------+--------+
              0 |     20 |     13 |     33
      ---------+--------+--------+
              1 |      3 |     17 |     20
      ---------+--------+--------+
      Total          23        30       53
```

Explanation

In Output 4.3 you can see that only 3 of the observed events are incorrectly classified as predicted nonevents, but 13 observed nonevents are incorrectly classified as events. Also, the overall predictive accuracy of the logistic regression model is lower than it was for the probability cutpoint of .50. Only (20 + 17) = 37, or 70 percent of the observations are correctly classified by the model using the probability cutpoint of .25. Set the probability cutpoint according to whether the number of false negatives, the number of false positives, or the overall accuracy of the predictive model is most important to you.

Classifying New Observations

To test the predictive accuracy of your logistic regression model on a different set of observations than those data used to fit the model, you need to create a SAS data set with missing values for the response variable. Then, you merge the new data with the original data and run the logistic regression on the merged data. Because the new observations have missing values for the response variable, they are not used to fit the logistic regression model. However, the output data set contains predicted probabilities for all observations in the input data set, including the observations with missing values for the response variable. If you know the observed responses for the new observations, then you can test the predictive accuracy of the model just as you did before.

Program (continued)

Create a new data set, NEW1, that contains 10 new observations of prostate data.

```
data new1;
   input acid xray size nodalinv;
   lacd=log(acid);
   nodeinv=nodalinv;
   nodalinv=.;
   datalines;
   .42 1 0 0
   .66 1 0 0
   .85 0 1 1
   .52 1 0 0
   .61 0 0 0
   .75 0 0 1
   .67 0 0 0
   .67 1 1 1
   .49 0 0 0
   .45 0 0 0
;
```

Combine the new data set with the original prostate data.

```
data prostat2;
   set prostate new1;
run;
```

Fit the logistic regression model to the data and create an output data set that contains predicted probabilities.

```
proc logistic data=prostat2 descending noprint;
   model nodalinv=lacd xray size;
   output out=probs3 predicted=phat;
run;
```

List the relevant variables from the output data set.

```
proc print data=probs3;
   var lacd xray size nodalinv nodeinv phat;
run;
```

Output: PROC PRINT

Output 4.4
Listing of Some Variables from the Output Data Set from PROC LOGISTIC

Prostate Data

OBS	LACD	XRAY	SIZE	NODALINV	NODEINV	PHAT
1	-0.73397	0	0	0	.	0.05306
2	-0.57982	0	0	0	.	0.07389
3	-0.69315	0	0	0	.	0.05796
4	-0.65393	0	0	0	.	0.06307
5	-0.69315	0	0	0	.	0.05796
6	-0.71335	0	0	0	.	0.05549
7	-0.77653	1	0	0	.	0.28407
8	-0.47804	1	0	0	.	0.44025
9	-0.57982	0	0	1	.	0.07389
10	-0.59784	1	0	0	.	0.37408
11	-0.47804	0	0	0	.	0.09153
12	-0.34249	0	0	0	.	0.12085
13	-0.43078	0	0	0	.	0.10094
14	-0.40048	1	0	1	.	0.48442
15	-0.75502	0	0	0	.	0.05069
16	-0.71335	0	0	0	.	0.05549
17	-0.69315	0	0	0	.	0.05796
18	-0.24846	0	0	0	.	0.14568
19	-0.18633	0	0	0	.	0.16431
20	-0.02020	0	0	0	.	0.22345
21	-0.65393	0	0	0	.	0.06307
22	-0.28768	0	0	0	.	0.13484
23	-0.01005	0	0	1	.	0.22751
24	0.62594	0	0	0	.	0.55859
25	0.30748	1	0	1	.	0.82642
26	-0.19845	0	0	1	.	0.16053
27	-0.91629	0	1	0	.	0.17712
28	-0.69315	0	1	0	.	0.26416
29	-0.69315	0	1	0	.	0.26416
30	-0.91629	0	1	0	.	0.17712
31	-0.59784	0	1	0	.	0.30874
32	-0.52763	0	1	0	.	0.34410
33	-0.73397	1	1	1	.	0.71849
34	-0.67334	1	1	1	.	0.74572
35	-0.71335	0	1	1	.	0.25525
36	-0.73397	0	1	0	.	0.24637
37	-0.46204	1	1	0	.	0.82640
38	0.01980	0	1	0	.	0.64789
39	-0.27444	0	1	0	.	0.48383
40	-0.05129	0	1	0	.	0.60988
41	-0.41552	0	1	0	.	0.40418
42	-0.17435	1	1	1	.	0.90201
43	-0.21072	1	1	1	.	0.89439
44	-0.27444	1	1	1	.	0.87978
45	-0.35667	0	1	1	.	0.43704
46	-0.24846	1	1	1	.	0.88593
47	-0.35667	0	1	1	.	0.43704
48	-0.40048	0	1	1	.	0.41251
49	-0.19845	0	1	1	.	0.52735
50	-0.40048	0	1	1	.	0.41251
51	-0.32850	1	1	1	.	0.86604
52	-0.11653	1	1	1	.	0.91312
53	0.23111	1	1	1	.	0.95888
54	-0.86750	1	0	.	0	0.24363
55	-0.41552	1	0	.	0	0.47581
56	-0.16252	0	1	.	1	0.54782
57	-0.65393	1	0	.	0	0.34450
58	-0.49430	0	0	.	0	0.08847
59	-0.28768	0	0	.	1	0.13484
60	-0.40048	0	0	.	0	0.10742
61	-0.40048	1	1	.	1	0.84572
62	-0.71335	0	0	.	0	0.05549
63	-0.79851	0	0	.	0	0.04610

Explanation

Note that the new observations are listed at the bottom of Output 4.4. The NODALINV variable is missing for the 10 new observations. The NODEINV variable, which contains the observed responses for the new observations, is missing for the original 53 observations in the prostate data. The PHAT variable contains a predicted probability for all observations in the output data set.

Program (continued)

Revise the output data set from PROC LOGISTIC. Add a predicted response variable using a probability cutpoint of .50.

```
data probs4;
   set probs3;
   if phat>=.5 then predicts=1;
   else predicts=0;
run;
```

Create a classification table for the 10 new observations.

```
proc freq data=probs4;
   tables nodeinv*predicts / norow nocol nopercent;
run;
```

Output: Classification Table for New Data

Output 4.5
Classification Table for New Data

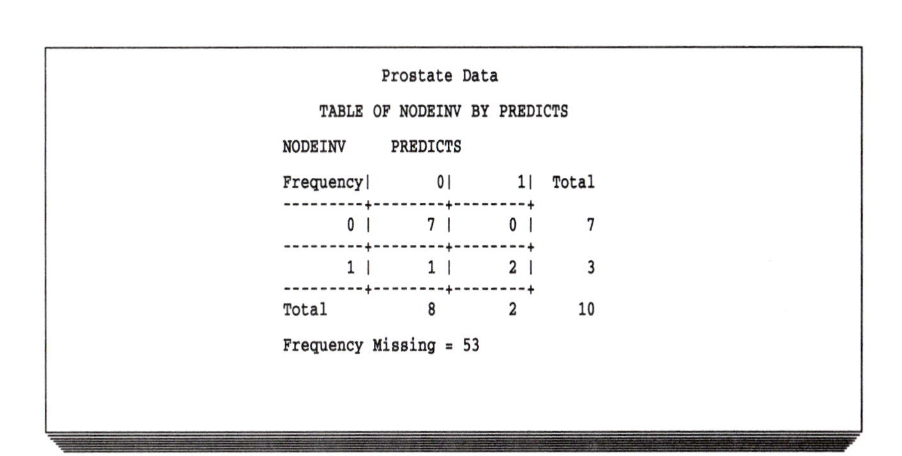

```
                    Prostate Data

             TABLE OF NODEINV BY PREDICTS

         NODEINV     PREDICTS

         Frequency|       0|       1|  Total
         ---------+--------+--------+
                0 |      7 |      0 |      7
         ---------+--------+--------+
                1 |      1 |      2 |      3
         ---------+--------+--------+
         Total           8        2       10

         Frequency Missing = 53
```

Explanation

In Output 4.5, you can see that (7 + 2) = 9, or 90 percent, of the 10 new observations are correctly classified. Only one observation is mistakenly classified as a nonevent. However, of the three observed events in these 10 new observations, only two are correctly classified as events. This sample size may be too small to give you a good idea of the predictive accuracy of your model.

The 53 observations from the original prostate data are excluded from the classification table because they have missing values for the NODEINV variable.

Program (continued)

Compute the Brier score for the 10 new observations.

```
data probs5;
   set probs3;
   brier=(phat-nodeinv)**2;
run;

proc means data=probs5 noprint;
   var brier;
   output out=brier1 n=n mean=;
run;

proc print data=brier1;
   var n brier;
   title2 'Brier Score';
run;
```

Output: Brier Score

Output 4.6
Brier Score for 10 New Observations

```
                Prostate Data
                 Brier Score

            OBS     N      BRIER

             1     10     0.14058
```

Explanation

The Brier score for the 10 new observations is .14. The range of the Brier score is 0 to 1. The smaller the score, the better the predictive ability of the model. Use the Brier score to compare the predictive ability of different models.

Further Reading

- For more information on DATA step processing, see *SAS Language: Reference, Version 6, First Edition.*
- For complete reference information on the FREQ and MEANS procedures, see the *SAS Procedures Guide, Version 6, Third Edition.*

References

- Bamber, D. (1975), "The Area Above the Ordinal Dominance Graph and the Area Below the Receiver Operating Characteristic Graph," *Journal of Mathematical Psychology*, 12, 387-415.
- Brier, G.W. (1950), "Verification of Forecasts Expressed in Terms of Probability," *Monthly Weather Review*, 75, 1-3.
- Hanley, J.A. and McNeil, B.J. (1982), "The Meaning and Use of the Area Under a Receiver Operating Characteristic (ROC) Curve," *Radiology*, 143, 29-36.

EXAMPLE 5

Creating Classification Tables

Featured Tools:
PROC LOGISTIC, MODEL statement:

- □ CTABLE option
- □ PPROB= option
- □ PEVENT= option

A *classification table* uses a logistic regression model to classify observations as events or nonevents. A classification table also measures the predictive accuracy of a logistic regression model, although it can be insensitive for this purpose. The LOGISTIC procedure produces a classification table that contains several measures of predictive accuracy for each probability cutpoint. That is, the model classifies an observation as an event if its estimated probability is greater than or equal to a given probability cutpoint. Otherwise, the observation is classified as a nonevent. As the probability cutpoints increase in value, the more likely that an observation is classified as a nonevent. The classification table reports how well these classifications match the observed event or nonevent status of each observation.

If you use the same data to test the predictive accuracy of your model that you use to fit the model, it can bias the results. Example 4 shows one way to avoid this bias — use a completely new set of observations to test the predictive accuracy of your model. Another way to avoid the bias is to fit a model that omits each observation one at a time and then to classify each observation as an event or a nonevent based on the model that omits the observation being classified. This method (known as *jackknifing*) can be expensive and time-consuming if the data set is large. PROC LOGISTIC creates a classification table using a method that approximates this unbiased jackknifing method. See page 1092 in the *SAS/STAT User's Guide, Version 6, Fourth Edition* for more details on this jackknifing method.

This example uses the CTABLE option in the MODEL statement of PROC LOGISTIC to create the bias-adjusted classification table. It shows how to use the PPROB= option to specify probability cutpoints for the classification table. This example also shows how to specify prior probabilities for the event of interest (using the PEVENT= option, which is available in Release 6.10 and later releases) and explains how this affects the classification table. If you do not specify a prior probability, PROC LOGISTIC uses the observed sample proportion of events as the prior probability. Specifying a prior probability is most useful when the sample is stratified on the response, that is, when you construct a sample by oversampling events. In that case, the observed sample proportion of events may be very different from the appropriate prior probability.

Program

Create the PROSTATE data set. The complete PROSTATE data set is in the Introduction.

```
data prostate;
   input case age acid xray size grade nodalinv @@;
   lacd=log(acid);
   data lines;
1 66  .48 0 0 0 0   2 68  .56 0 0 0 0   3 66  .50 0 0 0 0
more data lines
;
```

Create the default classification table. Use the CTABLE option in the MODEL statement.

```
proc logistic data=prostate descending;
   model nodalinv=lacd xray size / ctable;
   title 'Prostate Data';
run;
```

Specify probability cutpoints for the classification table. Use a range of probabilities in the PPROB= option. Enclose the range in parentheses.

```
proc logistic data=prostate descending;
   model nodalinv=lacd xray size / ctable
                                   pprob=(.05 to 1.0 by .05);
run;
```

Specify prior probabilities for the event. Use the PEVENT= option. PROC LOGISTIC produces a separate classification table for each prior probability that you specify.

```
proc logistic data=prostate descending;
   model nodalinv = lacd xray size / ctable
                                   pprob=(.05 to 1.0 by .05)
                                   pevent= .25 .50;
run;
```

Output

Output 5.1
Default Classification Table from PROC LOGISTIC (Partial Output)

Prostate Data

The LOGISTIC Procedure

Classification Table

	❶ Correct		❷ Incorrect		❸ Percentages				
Prob Level	Event	Non-Event	Event	Non-Event	Correct	Sensi-tivity	Speci-ficity	False POS	False NEG
0.040	20	0	33	0	37.7	100.0	0.0	62.3	.
0.060	19	7	26	1	49.1	95.0	21.2	57.8	12.5
0.080	19	10	23	1	54.7	95.0	30.3	54.8	9.1
0.100	19	11	22	1	56.6	95.0	33.3	53.7	8.3
0.120	18	12	21	2	56.6	90.0	36.4	53.8	14.3
0.140	18	13	20	2	58.5	90.0	39.4	52.6	13.3
0.160	18	15	18	2	62.3	90.0	45.5	50.0	11.8
0.180	17	16	17	3	62.3	85.0	48.5	50.0	15.8
0.200	17	18	15	3	66.0	85.0	54.5	46.9	14.3
0.220	16	18	15	4	64.2	80.0	54.5	48.4	18.2
0.240	16	18	15	4	64.2	80.0	54.5	48.4	18.2
0.260	16	19	14	4	66.0	80.0	57.6	46.7	17.4
0.280	16	20	13	4	67.9	80.0	60.6	44.8	16.7
0.300	16	22	11	4	71.7	80.0	66.7	40.7	15.4
0.320	16	22	11	4	71.7	80.0	66.7	40.7	15.4
0.340	16	23	10	4	73.6	80.0	69.7	38.5	14.8
0.360	16	24	9	4	75.5	80.0	72.7	36.0	14.3
0.380	14	25	8	6	73.6	70.0	75.8	36.4	19.4
0.400	13	25	8	7	71.7	65.0	75.8	38.1	21.9
0.420	11	25	8	9	67.9	55.0	75.8	42.1	26.5
0.440	11	26	7	9	69.8	55.0	78.8	38.9	25.7
0.460	11	27	6	9	71.7	55.0	81.8	35.3	25.0

Prob Level	Correct Event	Correct Non-Event	Incorrect Event	Incorrect Non-Event	Correct	Sensi-tivity	Speci-ficity	False POS	False NEG
0.480	11	27	6	9	71.7	55.0	81.8	35.3	25.0
0.500	10	27	6	10	69.8	50.0	81.8	37.5	27.0
0.520	10	29	4	10	73.6	50.0	87.9	28.6	25.6
0.540	10	29	4	10	73.6	50.0	87.9	28.6	25.6
0.560	10	29	4	10	73.6	50.0	87.9	28.6	25.6
0.580	10	29	4	10	73.6	50.0	87.9	28.6	25.6
0.600	10	29	4	10	73.6	50.0	87.9	28.6	25.6
0.620	10	29	4	10	73.6	50.0	87.9	28.6	25.6
0.640	10	29	4	10	73.6	50.0	87.9	28.6	25.6
0.660	10	29	4	10	73.6	50.0	87.9	28.6	25.6
0.680	9	30	3	11	73.6	45.0	90.9	25.0	26.8
0.700	9	30	3	11	73.6	45.0	90.9	25.0	26.8
0.720	8	30	3	12	71.7	40.0	90.9	27.3	28.6
0.740	8	31	2	12	73.6	40.0	93.9	20.0	27.9
0.760	8	31	2	12	73.6	40.0	93.9	20.0	27.9
0.780	8	31	2	12	73.6	40.0	93.9	20.0	27.9
0.800	7	31	2	13	71.7	35.0	93.9	22.2	29.5
0.820	7	31	2	13	71.7	35.0	93.9	22.2	29.5
0.840	7	32	1	13	73.6	35.0	97.0	12.5	28.9
0.860	6	32	1	14	71.7	30.0	97.0	14.3	30.4
0.880	4	32	1	16	67.9	20.0	97.0	20.0	33.3
0.900	2	33	0	18	66.0	10.0	100.0	0.0	35.3
0.920	1	33	0	19	64.2	5.0	100.0	0.0	36.5
0.940	1	33	0	19	64.2	5.0	100.0	0.0	36.5
0.960	0	33	0	20	62.3	0.0	100.0	.	37.7

Output 5.2
Classification Table with Specified Probability Cutpoints (Partial Output)

Prostate Data

The LOGISTIC Procedure

Classification Table

Prob Level	Correct Event	Correct Non-Event	Incorrect Event	Incorrect Non-Event	Correct	Sensi-tivity	Speci-ficity	False POS	False NEG
0.050	19	0	33	1	35.8	95.0	0.0	63.5	100.0
0.100	19	11	22	1	56.6	95.0	33.3	53.7	8.3
0.150	18	14	19	2	60.4	90.0	42.4	51.4	12.5
0.200	17	18	15	3	66.0	85.0	54.5	46.9	14.3
0.250	16	19	14	4	66.0	80.0	57.6	46.7	17.4
0.300	16	22	11	4	71.7	80.0	66.7	40.7	15.4
0.350	16	24	9	4	75.5	80.0	72.7	36.0	14.3
0.400	13	25	8	7	71.7	65.0	75.8	38.1	21.9
0.450	11	27	6	9	71.7	55.0	81.8	35.3	25.0
0.500	10	27	6	10	69.8	50.0	81.8	37.5	27.0
0.550	10	29	4	10	73.6	50.0	87.9	28.6	25.6
0.600	10	29	4	10	73.6	50.0	87.9	28.6	25.6
0.650	10	29	4	10	73.6	50.0	87.9	28.6	25.6
0.700	9	30	3	11	73.6	45.0	90.9	25.0	26.8
0.750	8	31	2	12	73.6	40.0	93.9	20.0	27.9
0.800	7	31	2	13	71.7	35.0	93.9	22.2	29.5
0.850	7	32	1	13	73.6	35.0	97.0	12.5	28.9
0.900	2	33	0	18	66.0	10.0	100.0	0.0	35.3
0.950	1	33	0	19	64.2	5.0	100.0	0.0	36.5
1.000	0	33	0	20	62.3	0.0	100.0	.	37.7

Output 5.3
Classification Tables for Specified Prior Probabilities (Partial Output)

Prostate Data

The LOGISTIC Procedure

Classification Table

Prob Event	Prob Level	Correct Event	Correct Non-Event	Incorrect Event	Incorrect Non-Event	Correct	Sensi-tivity	Speci-ficity	False POS	False NEG
0.250	0.050	19	0	33	1	23.8	95.0	0.0	75.9	100.0
0.250	0.100	19	11	22	1	48.8	95.0	33.3	67.8	4.8
0.250	0.150	18	14	19	2	54.3	90.0	42.4	65.7	7.3
0.250	0.200	17	18	15	3	62.2	85.0	54.5	61.6	8.4

```
0.250 0.250   16   19   14    4   63.2  80.0   57.6  61.4  10.4
0.250 0.300   16   22   11    4   70.0  80.0   66.7  55.6   9.1
0.250 0.350   16   24    9    4   74.5  80.0   72.7  50.6   8.4
0.250 0.400   13   25    8    7   73.1  65.0   75.8  52.8  13.3
0.250 0.450   11   27    6    9   75.1  55.0   81.8  49.8  15.5
0.250 0.500   10   27    6   10   73.9  50.0   81.8  52.2  16.9
0.250 0.550   10   29    4   10   78.4  50.0   87.9  42.1  15.9
0.250 0.600   10   29    4   10   78.4  50.0   87.9  42.1  15.9
0.250 0.650   10   29    4   10   78.4  50.0   87.9  42.1  15.9
0.250 0.700    9   30    3   11   79.4  45.0   90.9  37.7  16.8
0.250 0.750    8   31    2   12   80.5  40.0   93.9  31.3  17.6
0.250 0.800    7   31    2   13   79.2  35.0   93.9  34.2  18.7
0.250 0.850    7   32    1   13   81.5  35.0   97.0  20.6  18.3
0.250 0.900    2   33    0   18   77.5  10.0  100.0   0.0  23.1
0.250 0.950    1   33    0   19   76.3   5.0  100.0   0.0  24.1
0.250 1.000    0   33    0   20   75.0   0.0  100.0     .  25.0

0.500 0.050   19    0   33    1   47.5  95.0    0.0  51.3 100.0
0.500 0.100   19   11   22    1   64.2  95.0   33.3  41.2  13.0
0.500 0.150   18   14   19    2   66.2  90.0   42.4  39.0  19.1
0.500 0.200   17   18   15    3   69.8  85.0   54.5  34.8  21.6
0.500 0.250   16   19   14    4   68.8  80.0   57.6  34.7  25.8
0.500 0.300   16   22   11    4   73.3  80.0   66.7  29.4  23.1
0.500 0.350   16   24    9    4   76.4  80.0   72.7  25.4  21.6
0.500 0.400   13   25    8    7   70.4  65.0   75.8  27.2  31.6
0.500 0.450   11   27    6    9   68.4  55.0   81.8  24.8  35.5
0.500 0.500   10   27    6   10   65.9  50.0   81.8  26.7  37.9
0.500 0.550   10   29    4   10   68.9  50.0   87.9  19.5  36.3
0.500 0.600   10   29    4   10   68.9  50.0   87.9  19.5  36.3
0.500 0.650   10   29    4   10   68.9  50.0   87.9  19.5  36.3
0.500 0.700    9   30    3   11   68.0  45.0   90.9  16.8  37.7
0.500 0.750    8   31    2   12   67.0  40.0   93.9  13.2  39.0
0.500 0.800    7   31    2   13   64.5  35.0   93.9  14.8  40.9
0.500 0.850    7   32    1   13   66.0  35.0   97.0   8.0  40.1
0.500 0.900    2   33    0   18   55.0  10.0  100.0   0.0  47.4
0.500 0.950    1   33    0   19   52.5   5.0  100.0   0.0  48.7
0.500 1.000    0   33    0   20   50.0   0.0  100.0     .  50.0
```

Explanation

Output 5.1 shows the default classification table that PROC LOGISTIC creates when you specify the CTABLE option. It lists the classification for a range of probabilities from the smallest estimated probability (rounded down to the nearest 0.02) to the highest estimated probability (rounded up to the nearest 0.02) with 0.02 increments. For the prostate data, the lowest estimated probability is about 0.05 and the highest estimated probability is about 0.96. See Example 4 for more information on the fit of the model to the prostate data.

CAUTION!

> When you select a "best" classification rule from a range of probability cutpoints in a classification table, that rule may not validate with new data. That is, classification tables are subject to the bias of multiple comparisons. ■

The columns labeled `Correct` ❶ and `Incorrect` ❷ give the frequency with which observations are correctly and incorrectly, respectively, classified as events or nonevents for each probability cutpoint. For example, at the cutpoint of 0.06, the model correctly classifies 19 events and 7 nonevents. It incorrectly classifies 26 nonevents as events and 1 event as a nonevent.

The five `Percentages` ❸ columns measure the predictive accuracy of the model:

`Correct`
: gives the probability that the model correctly classifies the sample data for each probability cutpoint. If you do not specify a prior probability for the event, then the percentage correct is simply a ratio consisting of the number of correctly classified observations over the total number of observations. For the example of a 0.06 cutpoint, 26 of the 53 observations are correctly classified, which produces a percentage correct of 49.1 percent (26 / 53 = 49.1%).

`Sensitivity`
: is a ratio consisting of the number of correctly classified events over the total number of events. At the cutpoint of .06, the sensitivity is 95 percent because 19 of the 20 events are correctly classified (19 / 20 = 95%).

`Specificity`
: is a ratio consisting of the number of correctly classified nonevents over the total number of nonevents. At the cutpoint of 0.06, specificity is 21.2 percent because 7 of the 33 nonevents are correctly classified (7 / 33 = 21.2%).

`False POS`
: is the false positive rate. The computation of the false positive rate depends on whether or not you specify a prior probability for the event. If you do not specify a prior probability, then the false positive rate is a ratio consisting of the number of nonevents incorrectly classified as events over the sum of all observations classified as events. For the cutpoint of 0.06, 26 nonevents are incorrectly classified as events. A total of 45 observations are classified as events (19 + 26 = 45). So, the false positive rate is 57.8 percent (26 / 45 = 57.8%).

`False NEG`
: is the false negative rate. The computation of the false negative rate depends on whether or not you specify a prior probability for the event. If you do not specify a prior probability, then the false negative rate is a ratio consisting of the number of events incorrectly classified as nonevents over the sum of all observations classified as nonevents. For the cutpoint of 0.06, one event is incorrectly classified as a nonevent. A total of eight observations are classified as nonevents (7 + 1 = 8). So, the false negative rate is 12.5 percent (1 / 8 = 12.5%).

Compare the results of the biased classification table shown in Example 4 (Output 4.2) to the unbiased classification table in Output 5.1. For a cutpoint of 0.50 (the shaded line in Output 5.1), the percentage correct in the unbiased table is 69.8 percent. The table in Output 4.2 produced a percentage correct of 75 percent, which is optimistically biased.

Output 5.2 shows the classification table that PROC LOGISTIC creates when you specify a range of probability cutpoints with the PPROB= option. The cutpoints range from 0.05 to 1.0 with increments of 0.05. When you specify the PPROB= option, it has no effect on the way PROC LOGISTIC computes the measures of predictive accuracy in the classification table.

Output 5.3 shows two classification tables — one for a prior probability of 0.25 and one for a prior probability of 0.50. PROC LOGISTIC creates a separate classification table for each prior probability that you specify in the

PEVENT= option. By default, PROC LOGISTIC uses a prior probability equal to the sample proportion of events. For the prostate data, this sample proportion is about 0.377. The prior probability affects the computations in three of the **`Percentages`** columns: **`Correct`**, **`False POS`**, and **`False NEG`**. When you specify a prior probability, then the false positive and false negative rates are computed as posterior probabilities with Bayes rule. For details on how Bayes rule is applied to compute the false positive and false negative rates, see pages 50-51 in *SAS/STAT Changes and Enhancements, Release 6.10*.

The percentage correct is computed as follows:

$$\% \text{ Correct} = (p \times \text{Sensitivity}) + [(1 - p) \times \text{Specificity}]$$

where p is the prior probability.

Reference

□ SAS Institute Inc. (1994), *SAS/STAT Changes and Enhancements, Release 6.10*, Cary, NC: SAS Institute Inc.

EXAMPLE 6

Using Model Selection Methods in Logistic Regression

Featured Tools:
PROC LOGISTIC, MODEL statement:

- BEST= option
- DETAILS option
- INCLUDE= option
- SELECTION= option
- SEQUENTIAL option
- SLENTRY= option
- SLSTAY= option
- START= option
- STOP= option

Model selection is the process of adding variables to or removing variables from a model until you find the model that is relatively the best of the competing models for the data. At each stage in the process, the significance levels of the variables are computed and compared to specified significance level criteria for entry to or removal from the model.

Several model selection methods are available:

Forward selection
: adds variables one at a time to the model as those variables meet the specified significance level for entry to the model.

Backward elimination
: removes variables one at a time from the model as those variables fail to meet the specified significance level for staying in the model.

Stepwise selection
: combines the backward elimination and forward selection methods to add variables to the model or remove variables from the model as they meet or fail to meet specified significance levels, respectively.

Best subset selection
: finds a specified number of models with the highest likelihood score (chi-square) statistic for all possible model sizes, from one, two, three variables, and so on, up to the single model that contains all of the explanatory variables.

Use model selection methods in logistic regression with care. You capitalize on chance by comparing significance levels for the different combinations of explanatory variables. When you perform many significance tests, each at a level of, say 5 percent, the overall probability of rejecting at least one true null hypothesis is much larger than 5 percent. One way to deal with this problem is to specify a very small significance level for each test that you perform. Furthermore, none of the *p*-values for the parameter estimates have the conventional meaning because none of the test statistics has a normal or chi-square distribution. Also, take care not to consider more candidate variables than can have parameters reliably estimated from your sample.

Model selection techniques are exploratory. It is useful to verify the fit of the selected model on other data.

This example uses options in the MODEL statement of the LOGISTIC procedure in several programs to demonstrate various model selection methods.

Program: Forward Selection

Create the PROSTATE data set. The complete PROSTATE data set is in the Introduction.

```
data prostate;
   input case age acid xray size grade nodalinv @@;
   lacd=log(acid);
   datalines;
1  66  .48 0 0 0 0   2  68  .56 0 0 0 0   3  66  .50 0 0 0 0
more data lines
;
```

Fit a forward stepwise logistic regression model. Set the entry criterion of .25 with the SLENTRY= option.
■ The DETAILS option provides information on the model selection process.

```
proc logistic data=prostate descending;
   model nodalinv=age lacd xray size grade / selection=forward
                                             slentry=.25
                                             details;
   title 'Prostate Data';
run;
```

Specify that 2 is the maximum number of explanatory variables that can be added to the model with forward selection. Use the STOP= option.

```
proc logistic data=prostate descending;
   model nodalinv=age lacd xray size grade / selection=forward
                                             slentry=.25
                                             details
                                             stop=2;
run;
```

Output

Output: Forward Selection

Output 6.1
Forward Selection Using PROC LOGISTIC (Partial Output)

```
                              Prostate Data
                         The LOGISTIC Procedure
                       Forward Selection Procedure

   Step  0. Intercept entered:

                 ❶ Analysis of Maximum Likelihood Estimates

                Parameter Standard    Wald       Pr >   Standardized     Odds
   Variable DF  Estimate   Error   Chi-Square Chi-Square  Estimate      Ratio

   INTERCPT 1    -0.5008   0.2834     3.1229     0.0772            .       .
```

```
❷ Residual Chi-Square = 20.4784 with 5 DF (p=0.0010)

      ❸ Analysis of Variables Not in the Model

                       Score        Pr >
       Variable      Chi-Square   Chi-Square

       AGE             1.0936       0.2957
       LACD            5.3243       0.0210
       XRAY           11.2831       0.0008
       SIZE            7.4383       0.0064
       GRADE           4.0746       0.0435

Step  1. Variable XRAY entered:

      ❹ Testing Global Null Hypothesis: BETA=0

                             Intercept
              Intercept         and
Criterion       Only        Covariates    Chi-Square for Covariates

AIC            72.252         63.001          .
SC             74.222         66.941          .
-2 LOG L       70.252         59.001       11.251 with 1 DF (p=0.0008)
Score             .              .         11.283 with 1 DF (p=0.0008)

               Analysis of Maximum Likelihood Estimates

             Parameter Standard    Wald       Pr >    Standardized     Odds
Variable DF  Estimate    Error  Chi-Square Chi-Square   Estimate      Ratio

INTERCPT 1    -1.1701   0.3816     9.4033    0.0022          .          .
XRAY     1     2.1817   0.6975     9.7835    0.0018     0.547014     8.861

❺ Association of Predicted Probabilities and Observed Responses

          Concordant = 48.3%          Somers' D = 0.429
          Discordant =  5.5%          Gamma     = 0.797
          Tied       = 46.2%          Tau-a     = 0.205
          (660 pairs)                 c         = 0.714

     Residual Chi-Square = 11.6479 with 4 DF (p=0.0202)

          Analysis of Variables Not in the Model

                       Score        Pr >
       Variable      Chi-Square   Chi-Square

       AGE             1.3523       0.2449
       LACD            3.7971       0.0513
       SIZE            5.6394       0.0176
       GRADE           2.3710       0.1236

Step  2. Variable SIZE entered:

         Testing Global Null Hypothesis: BETA=0

                             Intercept
              Intercept         and
Criterion       Only        Covariates    Chi-Square for Covariates

AIC            72.252         59.353          .
SC             74.222         65.264          .
-2 LOG L       70.252         53.353       16.899 with 2 DF (p=0.0002)
Score             .              .         15.714 with 2 DF (p=0.0004)

               Analysis of Maximum Likelihood Estimates

             Parameter Standard    Wald       Pr >    Standardized     Odds
Variable DF  Estimate    Error  Chi-Square Chi-Square   Estimate      Ratio

INTERCPT 1    -2.0446   0.6100    11.2360    0.0008          .          .
XRAY     1     2.1194   0.7468     8.0537    0.0045     0.531411     8.326
SIZE     1     1.5883   0.7000     5.1479    0.0233     0.441942     4.895

  Association of Predicted Probabilities and Observed Responses

          Concordant = 68.8%          Somers' D = 0.580
          Discordant = 10.8%          Gamma     = 0.730
          Tied       = 20.5%          Tau-a     = 0.278
          (660 pairs)                 c         = 0.790

     Residual Chi-Square = 6.6755 with 3 DF (p=0.0830)
```

```
                     Analysis of Variables Not in the Model

                                     Score           Pr >
                     Variable      Chi-Square     Chi-Square

                     AGE             1.2678         0.2602
                     LACD            4.5808         0.0323
                     GRADE           0.5839         0.4448

Step  3. Variable LACD entered:

                    Testing Global Null Hypothesis: BETA=0

                                     Intercept
                    Intercept           and
Criterion             Only          Covariates    Chi-Square for Covariates

AIC                  72.252           56.986            .
SC                   74.222           64.867            .
-2 LOG L             70.252           48.986       21.266 with 3 DF (p=0.0001)
Score                   .                .         18.893 with 3 DF (p=0.0003)

                    Analysis of Maximum Likelihood Estimates

               Parameter Standard    Wald       Pr >    Standardized     Odds
Variable DF    Estimate   Error   Chi-Square Chi-Square   Estimate       Ratio

INTERCPT 1     -1.1994   0.7162     2.8046     0.0940          .           .
LACD     1      2.2922   1.1387     4.0520     0.0441      0.398208     9.897
XRAY     1      2.0550   0.7976     6.6380     0.0100      0.515257     7.807
SIZE     1      1.7638   0.7483     5.5562     0.0184      0.490771     5.834

        Association of Predicted Probabilities and Observed Responses

                  Concordant = 84.2%           Somers' D = 0.688
                  Discordant = 15.5%           Gamma     = 0.690
                  Tied       =  0.3%           Tau-a     = 0.329
                  (660 pairs)                  c         = 0.844

              Residual Chi-Square = 2.4067 with 2 DF (p=0.3002)

                     Analysis of Variables Not in the Model

                                     Score           Pr >
                     Variable      Chi-Square     Chi-Square

                     AGE             1.3006         0.2541
                     GRADE           1.2319         0.2670

❻  NOTE: No (additional) variables met the 0.25 significance level for entry
          into the model.

                  ❼  Summary of Forward Selection Procedure

                     Variable     Number        Score          Pr >
              Step   Entered        In        Chi-Square     Chi-Square

                1    XRAY           1          11.2831         0.0008
                2    SIZE           2           5.6394         0.0176
                3    LACD           3           4.5808         0.0323
```

Explanation

Output 6.1 shows the forward selection process in PROC LOGISTIC:

□ At `Step 0`, the intercept is entered into the model. PROC LOGISTIC prints the **`Analysis of Maximum Likelihood Estimates`** table ❶, the residual chi-square statistic ❷, and the **`Analysis of Variables Not in the Model`** table ❸. The *p*-value of .001 for the residual chi-square indicates that at least one of the remaining parameter coefficients is nonzero. That is, at least one parameter coefficient for a variable currently not in the model is nonzero. From the table of remaining variables, note that the variable XRAY has the largest score statistic and a *p*-value of

0.0008. This *p*-value is less than the criterion of 0.25 for entry into the model.

- At **Step 1**, the variable XRAY is entered into the model. PROC LOGISTIC prints the **Testing Global Null Hypothesis: BETA=0** table ❹, which contains the AIC, SC, -2 LOG L, and Score statistics. Also, PROC LOGISTIC prints the **Association of Predicted Probabilities and Observed Responses** table ❺, along with the same tables and statistics as in **Step 0**. In the **Analysis of Variables Not in the Model** table, note that the variable SIZE has the largest score statistic, with a *p*-value of 0.0176. This *p*-value is less than the criterion of 0.25 for entry into the model.
- At **Step 2**, the variable SIZE is entered into the model, and PROC LOGISTIC prints the same statistics and tables as in **Step 1**. In the **Analysis of Variables Not in the Model** table, note that the variable LACD has the largest score statistic, with a *p*-value of 0.0323. This *p*-value is less than the criterion of 0.25 for entry into the model.
- At **Step 3**, the variable LACD is entered into the model, and PROC LOGISTIC prints the same statistics and tables as in **Step 1** and **Step 2**. In the **Analysis of Variables Not in the Model** table, note that neither of the remaining two variables have *p*-values less than the criterion of 0.25 for entry into the model.

When the selection process is complete, a note states that no additional variables met the specified significance level for entry into the model ❻. This is the significance level that you specify with the SLENTRY= option.

A table in the output summarizes the forward selection procedure ❼. This table lists step number, the name of each explanatory variable that is entered into the model at each step, the chi-square statistic for each variable, and the corresponding probability value upon which each variable's entry into the model is based.

The final model contains an intercept and the XRAY, SIZE, and LACD variables.

Output: Using the STOP= Option

Output 6.2
Forward Selection in PROC LOGISTIC Using the STOP= Option

```
                         Prostate Data
                    The LOGISTIC Procedure
                  Forward Selection Procedure

Step  0. Intercept entered:

              Analysis of Maximum Likelihood Estimates

           Parameter Standard    Wald       Pr >   Standardized     Odds
Variable DF Estimate   Error  Chi-Square Chi-Square   Estimate      Ratio

INTERCPT 1   -0.5008  0.2834    3.1229     0.0772            .         .

           Residual Chi-Square = 20.4784 with 5 DF (p=0.0010)
```

```
                  Analysis of Variables Not in the Model

                                     Score          Pr >
                  Variable        Chi-Square     Chi-Square

                  AGE                 1.0936         0.2957
                  LACD                5.3243         0.0210
                  XRAY               11.2831         0.0008
                  SIZE                7.4383         0.0064
                  GRADE               4.0746         0.0435

Step  1. Variable XRAY entered:

                 Testing Global Null Hypothesis: BETA=0

                                 Intercept
                 Intercept          and
Criterion          Only         Covariates    Chi-Square for Covariates

AIC               72.252         63.001            .
SC                74.222         66.941            .
-2 LOG L          70.252         59.001          11.251 with 1 DF (p=0.0008)
Score               .              .             11.283 with 1 DF (p=0.0008)

                 Analysis of Maximum Likelihood Estimates

             Parameter Standard    Wald        Pr >     Standardized     Odds
Variable DF  Estimate    Error  Chi-Square Chi-Square    Estimate       Ratio

INTERCPT 1    -1.1701   0.3816     9.4033     0.0022             .          .
XRAY     1     2.1817   0.6975     9.7835     0.0018      0.547014      8.861


      Association of Predicted Probabilities and Observed Responses

                Concordant = 48.3%          Somers' D = 0.429
                Discordant =  5.5%          Gamma     = 0.797
                Tied       = 46.2%          Tau-a     = 0.205
                (660 pairs)                 c         = 0.714

            Residual Chi-Square = 11.6479 with 4 DF (p=0.0202)

                  Analysis of Variables Not in the Model

                                     Score          Pr >
                  Variable        Chi-Square     Chi-Square

                  AGE                 1.3523         0.2449
                  LACD                3.7971         0.0513
                  SIZE                5.6394         0.0176
                  GRADE               2.3710         0.1236

Step  2. Variable SIZE entered:

                 Testing Global Null Hypothesis: BETA=0

                                 Intercept
                 Intercept          and
Criterion          Only         Covariates    Chi-Square for Covariates

AIC               72.252         59.353            .
SC                74.222         65.264            .
-2 LOG L          70.252         53.353          16.899 with 2 DF (p=0.0002)
Score               .              .             15.714 with 2 DF (p=0.0004)

                 Analysis of Maximum Likelihood Estimates

             Parameter Standard    Wald        Pr >     Standardized     Odds
Variable DF  Estimate    Error  Chi-Square Chi-Square    Estimate       Ratio

INTERCPT 1    -2.0446   0.6100    11.2360     0.0008             .          .
XRAY     1     2.1194   0.7468     8.0537     0.0045      0.531411      8.326
SIZE     1     1.5883   0.7000     5.1479     0.0233      0.441942      4.895


      Association of Predicted Probabilities and Observed Responses

                Concordant = 68.8%          Somers' D = 0.580
                Discordant = 10.8%          Gamma     = 0.730
                Tied       = 20.5%          Tau-a     = 0.278
                (660 pairs)                 c         = 0.790

8 NOTE: The number of explanatory variables in the model has reached STOP=2.
```

```
                Summary of Forward Selection Procedure

                Variable    Number       Score        Pr >
        Step    Entered         In  Chi-Square  Chi-Square

           1    XRAY             1     11.2831      0.0008
           2    SIZE             2      5.6394      0.0176
```

Explanation

Output 6.2 shows the effect of using the STOP= option with forward selection. PROC LOGISTIC follows the same steps as it followed in Output 6.1, but it stops after `Step 2` in this case because you specify that only two explanatory variables can be added to the model. The note states that the number of explanatory variables in the model has reached the specified STOP= value ❽. Only the XRAY and SIZE variables are entered into the model along with the intercept.

Program: Backward Elimination

Fit a backward stepwise logistic regression model using the BACKWARD option. Set the retention criterion of 0.10 with the SLSTAY= option.

```
proc logistic data=prostate descending;
   model nodalinv=age lacd xray size grade / selection=backward
                                              details
                                              slstay=.10;
run;
```

Output: Backward Elimination

Output 6.3
Backward Elimination Using PROC LOGISTIC (Partial Output)

```
                              Prostate Data
                          The LOGISTIC Procedure
                      Backward Elimination Procedure

Step  0. The following variables were entered:

       INTERCPT  AGE       LACD      XRAY      SIZE      GRADE

                 Testing Global Null Hypothesis: BETA=0

                             Intercept
                Intercept       and
Criterion         Only      Covariates    Chi-Square for Covariates

AIC              72.252       58.560            .
SC               74.222       70.382            .
-2 LOG L         70.252       46.560         23.692 with 5 DF (p=0.0002)
Score               .            .           20.478 with 5 DF (p=0.0010)

                 Analysis of Maximum Likelihood Estimates

           Parameter Standard    Wald        Pr >    Standardized     Odds
Variable DF  Estimate   Error Chi-Square Chi-Square    Estimate      Ratio

INTERCPT 1     2.4598  3.5222    0.4877    0.4849           .          .
AGE      1    -0.0637  0.0587    1.1763    0.2781     -0.216632      0.938
LACD     1     2.5725  1.1970    4.6188    0.0316      0.446893     13.098
XRAY     1     2.0401  0.8288    6.0583    0.0138      0.511515      7.691
SIZE     1     1.5466  0.7811    3.9205    0.0477      0.430357      4.696
GRADE    1     0.8345  0.7889    1.1188 ❶  0.2902 ❷    0.225142      2.304

          Association of Predicted Probabilities and Observed Responses

                Concordant = 86.5%          Somers' D = 0.730
                Discordant = 13.5%          Gamma     = 0.730
```

```
                Tied        =  0.0%           Tau-a     = 0.350
                (660 pairs)                   c         = 0.865


Step  1. Variable GRADE is removed:

                 Testing Global Null Hypothesis: BETA=0

                                 Intercept
                 Intercept          and
Criterion          Only         Covariates    Chi-Square for Covariates

AIC               72.252          57.681           .
SC                74.222          67.532           .
-2 LOG L          70.252          47.681        22.571 with 4 DF (p=0.0002)
Score               .               .           19.807 with 4 DF (p=0.0005)

                 Analysis of Maximum Likelihood Estimates

             Parameter Standard    Wald       Pr >    Standardized     Odds
Variable DF  Estimate   Error  Chi-Square Chi-Square   Estimate       Ratio

INTERCPT 1     2.6564   3.4902     0.5793     0.4466           .          .
AGE      1    -0.0652   0.0582     1.2585 ❸  0.2619 ❹ -0.221845     0.937
LACD     1     2.3494   1.1563     4.1285     0.0422      0.408138    10.479
XRAY     1     2.0900   0.8112     6.6380     0.0100      0.524017     8.085
SIZE     1     1.7565   0.7528     5.4451     0.0196      0.488757     5.792

      Association of Predicted Probabilities and Observed Responses

                Concordant = 84.7%            Somers' D = 0.694
                Discordant = 15.3%            Gamma     = 0.694
                Tied       =  0.0%            Tau-a     = 0.332
                (660 pairs)                   c         = 0.847

           Residual Chi-Square = 1.1450 with 1 DF (p=0.2846)


Step  2. Variable AGE is removed:

                 Testing Global Null Hypothesis: BETA=0

                                 Intercept
                 Intercept          and
Criterion          Only         Covariates    Chi-Square for Covariates

AIC               72.252          56.986           .
SC                74.222          64.867           .
-2 LOG L          70.252          48.986        21.266 with 3 DF (p=0.0001)
Score               .               .           18.893 with 3 DF (p=0.0003)

                 Analysis of Maximum Likelihood Estimates

             Parameter Standard    Wald       Pr >    Standardized     Odds
Variable DF  Estimate   Error  Chi-Square Chi-Square   Estimate       Ratio
                                                ❺
INTERCPT 1    -1.1994   0.7162     2.8046     0.0940           .          .
LACD     1     2.2922   1.1387     4.0520     0.0441      0.398208     9.897
XRAY     1     2.0550   0.7976     6.6380     0.0100      0.515257     7.807
SIZE     1     1.7638   0.7483     5.5562     0.0184      0.490771     5.834


      Association of Predicted Probabilities and Observed Responses

                Concordant = 84.2%            Somers' D = 0.688
                Discordant = 15.5%            Gamma     = 0.690
                Tied       =  0.3%            Tau-a     = 0.329
                (660 pairs)                   c         = 0.844

           Residual Chi-Square = 2.4067 with 2 DF (p=0.3002)

❻ NOTE: No (additional) variables met the 0.1 significance level for removal
        from the model.

           ❼ Summary of Backward Elimination Procedure

                  Variable     Number       Wald        Pr >
           Step   Removed          In   Chi-Square   Chi-Square

             1    GRADE             4       1.1188       0.2902
             2    AGE               3       1.2585       0.2619
```

Explanation

Output 6.3 shows the backward elimination process in PROC LOGISTIC.

- □ At `Step 0`, the intercept and all five explanatory variables are entered into the model. PROC LOGISTIC prints the **`Testing Global Null Hypothesis: BETA=0`** table, the **`Analysis of Maximum Likelihood Estimates`** table, and the **`Association of Predicted Probabilities and Observed Responses`** table. From the **`Analysis of Maximum Likelihood Estimates`** table, note that the variable GRADE has the smallest Wald chi-square value (1.1188) ❶, and consequently, the largest *p*-value (0.2902) ❷. This *p*-value is larger than the criterion of 0.10 for retention in the model.
- □ At `Step 1`, the variable GRADE is removed from the model. PROC LOGISTIC prints the same tables as in `Step 0`, and it also prints the residual chi-square value. From the **`Analysis of Maximum Likelihood Estimates`** table, note that the variable AGE has the smallest Wald chi-square value (1.2585) ❸ and the largest *p*-value (0.2619) ❹. This *p*-value is larger than the criterion of 0.10 for retention in the model.
- □ At `Step 2`, the variable AGE is removed from the model, and PROC LOGISTIC prints the same tables and statistics as in `Step 1`. From the **`Analysis of Maximum Likelihood Estimates`** table, note that none of the variables have *p*-values greater than the criterion of 0.10 for retention in the model ❺.

When the selection process is complete, a note states that no additional variables met the specified significance level for removal from the model ❻. This is the significance level that you specify with the SLSTAY= option.

A table in the output summarizes the backward elimination procedure. ❼ This table lists the step number, the name of each explanatory variable that is removed from the model at each step, the chi-square statistic for each variable, and the corresponding probability value upon which each variable's removal from the model is based.

The final model contains an intercept and the LACD, XRAY, and SIZE variables. Note that the final models computed from forward selection (Output 6.1) and backward elimination (Output 6.3) are identical in this example. This is not always the case. For example, if you specify a relatively large value for the SLENTRY= option with forward selection, then the final model is more likely to contain more explanatory variables. Similarly, if you specify a relatively large value for the SLSTAY= option with backward elimination, then the final model is more likely to contain more explanatory variables.

Program: Stepwise Selection

Fit a stepwise logistic regression model using the STEPWISE option. Specify a relatively large entry criterion, but use the default retention criterion of 0.05.

```
proc logistic data=prostate descending;
   model nodalinv=age lacd xray size grade / selection=stepwise
                                             slentry=.3;
run;
```

Begin the stepwise selection process with the first 4 variables in the MODEL statement. Use the START= option with the STEPWISE option.

```
proc logistic data=prostate descending;
   model nodalinv=age lacd xray size grade / selection=stepwise
                                             start=4;
run;
```

Include the first 4 variables in the MODEL statement in every model. Use the INCLUDE= option with the STEPWISE option.

```
proc logistic data=prostate descending;
   model nodalinv=age lacd xray size grade / selection=stepwise
                                             include=4;
run;
```

Output: Stepwise Selection

Output 6.4
Stepwise Selection Using PROC LOGISTIC (Partial Output)

```
                              Prostate Data
                          The LOGISTIC Procedure
                       Stepwise Selection Procedure

   Step  0. Intercept entered:

           Residual Chi-Square = 20.4784 with 5 DF (p=0.0010)


   Step  1. Variable XRAY entered:

               Testing Global Null Hypothesis: BETA=0

                                Intercept
                   Intercept       and
   Criterion         Only       Covariates    Chi-Square for Covariates

   AIC              72.252        63.001           .
   SC               74.222        66.941           .
   -2 LOG L         70.252        59.001        11.251 with 1 DF (p=0.0008)
   Score              .             .           11.283 with 1 DF (p=0.0008)

           Residual Chi-Square = 11.6479 with 4 DF (p=0.0202)

   Step  2. Variable SIZE entered:

               Testing Global Null Hypothesis: BETA=0

                                Intercept
                   Intercept       and
   Criterion         Only       Covariates    Chi-Square for Covariates

   AIC              72.252        59.353           .
   SC               74.222        65.264           .
   -2 LOG L         70.252        53.353        16.899 with 2 DF (p=0.0002)
   Score              .             .           15.714 with 2 DF (p=0.0004)

           Residual Chi-Square = 6.6755 with 3 DF (p=0.0830)
```

```
Step  3. Variable LACD entered:

                 Testing Global Null Hypothesis: BETA=0

                                Intercept
                 Intercept         and
Criterion          Only        Covariates    Chi-Square for Covariates

AIC               72.252         56.986           .
SC                74.222         64.867           .
-2 LOG L          70.252         48.986         21.266 with 3 DF (p=0.0001)
Score               .              .            18.893 with 3 DF (p=0.0003)

            Residual Chi-Square = 2.4067 with 2 DF (p=0.3002)

Step  4. Variable AGE entered:

                 Testing Global Null Hypothesis: BETA=0

                                Intercept
                 Intercept         and
Criterion          Only        Covariates    Chi-Square for Covariates

AIC               72.252         57.681           .
SC                74.222         67.532           .
-2 LOG L          70.252         47.681         22.571 with 4 DF (p=0.0002)
Score               .              .            19.807 with 4 DF (p=0.0005)

Step  5. Variable AGE is removed:

                 Testing Global Null Hypothesis: BETA=0

                                Intercept
                 Intercept         and
Criterion          Only        Covariates    Chi-Square for Covariates

AIC               72.252         56.986           .
SC                74.222         64.867           .
-2 LOG L          70.252         48.986         21.266 with 3 DF (p=0.0001)
Score               .              .            18.893 with 3 DF (p=0.0003)

                  ❶  Summary of Stepwise Procedure

           Variable            Number     Score        Wald         Pr >
Step    Entered    Removed        In   Chi-Square   Chi-Square   Chi-Square

   1    XRAY                       1      11.2831            .       0.0008
   2    SIZE                       2       5.6394            .       0.0176
   3    LACD                       3       4.5808            .       0.0323
   4    AGE                        4       1.3006            .       0.2541
   5               AGE             3            .       1.2585       0.2619

            ❷  Analysis of Maximum Likelihood Estimates

            Parameter Standard    Wald       Pr >    Standardized     Odds
Variable DF  Estimate    Error  Chi-Square Chi-Square   Estimate      Ratio

INTERCPT 1    -1.1994   0.7162     2.8046     0.0940          .          .
LACD     1     2.2922   1.1387     4.0520     0.0441   0.398208      9.897
XRAY     1     2.0550   0.7976     6.6380     0.0100   0.515257      7.807
SIZE     1     1.7638   0.7483     5.5562     0.0184   0.490771      5.834
```

Output 6.5
Stepwise Selection in PROC LOGISTIC Using the START= Option

```
                          Prostate Data

                     The LOGISTIC Procedure

                  Stepwise Selection Procedure


Step  0. The following variables were entered:
        INTERCPT  AGE       LACD      XRAY      SIZE


            Testing Global Null Hypothesis: BETA=0

                            Intercept
                Intercept      and
Criterion         Only      Covariates    Chi-Square for Covariates

AIC              72.252       57.681          .
SC               74.222       67.532          .
-2 LOG L         70.252       47.681        22.571 with 4 DF (p=0.0002)
Score               .            .          19.807 with 4 DF (p=0.0005)


Step  1. Variable AGE is removed:


            Testing Global Null Hypothesis: BETA=0

                            Intercept
                Intercept      and
Criterion         Only      Covariates    Chi-Square for Covariates

AIC              72.252       56.986          .
SC               74.222       64.867          .
-2 LOG L         70.252       48.986        21.266 with 3 DF (p=0.0001)
Score               .            .          18.893 with 3 DF (p=0.0003)


        Residual Chi-Square = 2.4067 with 2 DF (p=0.3002)


NOTE: No (additional) variables met the 0.05 significance level for entry
      into the model.

               ❸ Summary of Stepwise Procedure

          Variable           Number     Score       Wald        Pr >
Step   Entered    Removed       In   Chi-Square  Chi-Square  Chi-Square

   1                AGE          3         .        1.2585      0.2619

          ❹ Analysis of Maximum Likelihood Estimates

             Parameter Standard    Wald       Pr >    Standardized    Odds
Variable DF  Estimate    Error  Chi-Square Chi-Square   Estimate      Ratio

INTERCPT 1    -1.1994   0.7162     2.8046     0.0940          .         .
LACD     1     2.2922   1.1387     4.0520     0.0441     0.398208    9.897
XRAY     1     2.0550   0.7976     6.6380     0.0100     0.515257    7.807
SIZE     1     1.7638   0.7483     5.5562     0.0184     0.490771    5.834
```

Output 6.6
Stepwise Selection in PROC LOGISTIC Using the INCLUDE= Option

```
                          Prostate Data

                     The LOGISTIC Procedure

                  Stepwise Selection Procedure

The following variables will be included in each model:

        INTERCPT  AGE       LACD      XRAY      SIZE


Step  0. The INCLUDE variables were entered.


            Testing Global Null Hypothesis: BETA=0

                            Intercept
                Intercept      and
Criterion         Only      Covariates    Chi-Square for Covariates

AIC              72.252       57.681          .
SC               74.222       67.532          .
```

```
-2 LOG L         70.252        47.681        22.571 with 4 DF (p=0.0002)
Score               .             .          19.807 with 4 DF (p=0.0005)

            Residual Chi-Square = 1.1450 with 1 DF (p=0.2846)

❺ NOTE: No (additional) variables met the 0.05 significance level for entry
        into the model.

             ❻ Analysis of Maximum Likelihood Estimates

             Parameter Standard    Wald      Pr >    Standardized    Odds
Variable DF  Estimate   Error  Chi-Square Chi-Square   Estimate      Ratio

INTERCPT 1     2.6564   3.4902     0.5793     0.4466        .          .
AGE      1    -0.0652   0.0582     1.2585     0.2619    -0.221845    0.937
LACD     1     2.3494   1.1563     4.1285     0.0422     0.408138   10.479
XRAY     1     2.0900   0.8112     6.6380     0.0100     0.524017    8.085
SIZE     1     1.7565   0.7528     5.4451     0.0196     0.488757    5.792
```

Explanation

Output 6.4 shows the stepwise selection process in PROC LOGISTIC. The first four steps are the same as using forward selection with an entry significance level of 0.3. The fifth step is the same as using backward elimination with the default significance level of 0.05 for retention in the model. The stepwise method ends when a variable that was entered in the previous step is removed from the model. Note that in `Step 4`, the variable AGE is added to the model, and in `Step 5`, it is removed from the model.

The summary table ❶ lists each step number, the names of the explanatory variables entered or removed from the model at each step, the chi-square statistic, and the corresponding probability value upon which the entry or removal of each variable is based.

The final model contains the LACD, XRAY, and SIZE variables, just as it did for the forward selection and backward elimination stepwise models in this example. The **`Analysis of Maximum Likelihood Estimates`** table describes the final model ❷.

Output 6.5 and Output 6.6 show the difference between the START= and INCLUDE= options. The START= option begins the selection process with the first *n* variables, but does not guarantee that the final model will include the first *n* variables listed in the MODEL statement. The INCLUDE= option does guarantee that the final model will include at least the first *n* variables listed in the MODEL statement. When you use the START= option, the final model may include fewer or more variables than you specify in the START= option. When you use the INCLUDE= option, the final model may include more variables than you specify in the INCLUDE= option, but it will not include fewer variables.

Output 6.5 shows that the START= option causes PROC LOGISTIC to start the first step (`Step 0`) of the stepwise selection process with the first four explanatory variables entered into the model. In the next step (`Step 1`), the AGE variable is removed from the model. The summary table ❸ shows the chi-square statistic and probability value for the AGE variable. As before, the final model, which is described in the **`Analysis of Maximum Likelihood Estimates`** table ❹, contains the LACD, XRAY, and SIZE variables.

Output 6.6 shows that the INCLUDE= option causes PROC LOGISTIC to start the first step (**Step 0**) of the stepwise selection process with the first four explanatory variables entered into the model, just as the START= option did. The note in the output states that no additional variables met the significance level for entry into the model ❺. Thus, the final model includes the four INCLUDE variables. Note that the final model, which is described in the **Analysis of Maximum Likelihood Estimates** table ❻, contains the AGE variable, which is not statistically significant at the 0.05 alpha level. There is no summary table for the stepwise procedure because no explanatory variables are added to or removed from the model.

Program: Best Subset Selection

Produce a listing of the likelihood score (chi-square) statistic for all possible model sizes. Specify SELECTION=SCORE. Use the BEST= option to specify the number of models of each size to list.

```
proc logistic data=prostate descending;
   model nodalinv = age lacd xray size grade / selection=score
                                               best=2;
run;
```

Output: Best Subset Selection

Output 6.7
Best Subset Selection in PROC LOGISTIC (Partial Output)

```
                         Prostate Data

                     The LOGISTIC Procedure

            Regression Models Selected by Score Criterion

   Number of       Score
   Variables       Value    Variables Included in Model

           1     11.2831    XRAY
           1      7.4383    SIZE
   -------------------------------
           2     15.7144    XRAY SIZE
           2     14.2718    LACD XRAY
   ------------------------------------
           3     18.8925    LACD XRAY SIZE
           3     16.6959    LACD XRAY GRADE
   ------------------------------------------
           4     19.8071    AGE LACD XRAY SIZE
           4     19.6371    LACD XRAY SIZE GRADE
   -----------------------------------------------
           5     20.4784    AGE LACD XRAY SIZE GRADE
   ----------------------------------------------------
```

Explanation

Output 6.7 lists the two best models, as measured by the score value, of all models with one, two, three, four, and five explanatory variables. (Only one five-variable model exists.) The best models have the highest score values. For example, the best three-variable model contains the LACD, XRAY, and SIZE variables.

A Closer Look

The DETAILS option

The DETAILS option in the MODEL statement produces a complete analysis of the model at each step of the selection process. This option produces a table called **`Analysis of Variables Not in the Model`** before printing the variable selected for entry for forward or stepwise selection. This table gives the score chi-square statistic for testing the significance of each variable not in the model after adjusting for the variables already in the model. The variable with the highest score statistic (with a *p*-value that meets the criterion for entry) is added to the model at the next step. This table also prints the probability value of the score chi-square with respect to a chi-square distribution with one degree of freedom.

For each model fitted, the DETAILS option also produces two other tables at each step.

- The **`Analysis of Maximum Likelihood Estimates`** table contains estimated parameters, standard errors, chi-square statistics, and *p*-values to help you assess the fit of the model at each step in the process. PROC LOGISTIC uses the *p*-value of the Wald chi-square in this table to remove a variable from the model when its *p*-value fails to meet the criterion for retention.
- The **`Association of Predicted Probabilities and Observed Responses`** table lists several measures of association to help you compare the predictive ability of the fitted model to other models. See Example 1 for more information about this table.

Variations

Using Other Model Selection Options

Specify the SEQUENTIAL option in the MODEL statement to force PROC LOGISTIC to add variables to the model in the order specified in the MODEL statement when you perform forward or stepwise selection. This option forces PROC LOGISTIC to remove variables from the model in the reverse order specified in the MODEL statement when you perform backward elimination.

To fit a complete model containing all the explanatory variables listed in the MODEL statement, specify SELECTION=NONE (or omit the SELECTION= option). The default for SELECTION= is NONE.

EXAMPLE 7

Computing Goodness-of-Fit Tests and Measures for Logistic Regression Models

Featured Tools:
PROC LOGISTIC, MODEL statement:

- □ LACKFIT option
- □ RSQ option

A formal statistical test for the goodness-of-fit of your logistic regression model gives you an objective measure of how well your model fits the data. This example discusses three measures of fit:

- □ the Hosmer and Lemeshow goodness-of-fit test
- □ a generalized coefficient of determination (R-square)
- □ an adjusted generalized coefficient of determination (R-square).

Hosmer and Lemeshow Goodness-of-Fit Test

Hosmer and Lemeshow (1989) developed a goodness-of-fit test for logistic regression models with binary responses. This test involves dividing the data into approximately ten groups of roughly equal size based on the percentiles of the estimated probabilities. The observations are sorted in increasing order of their estimated probability of having an event outcome. The discrepancies between the observed and expected number of observations in these groups are summarized by the Pearson chi-square statistic, which is then compared to a chi-square distribution with *t* degrees of freedom, where *t* is the number of groups minus 2.

The Hosmer and Lemeshow goodness-of-fit statistic is obtained by calculating the Pearson chi-square statistic from the 2 x *g* table of observed and expected frequencies, where *g* is the number of groups. The statistic is written

$$X^2_{HW} = \sum_{i=1}^{g} \frac{(O_i - N_i\bar{\pi}_i)^2}{N_i\bar{\pi}_i(1 - \bar{\pi}_i)}$$

where

N_i is the number of observations in the *i*th group.

O_i is the number of event outcomes in the *i*th group.

$\bar{\pi}_i$ is the average estimated probability of an event outcome for the *i*th group.

The Hosmer and Lemeshow statistic is then compared to a chi-square distribution with (*g* - 2) degrees of freedom.

Note these warnings about the Hosmer and Lemeshow goodness-of-fit test:

- □ It is a conservative test.
- □ It has low power to detect specific types of lack of fit (such as nonlinearity in an explanatory variable).

- It is highly dependent on how the observations are grouped.
- If too few groups are used to calculate the statistic (for example, five or fewer groups), it will almost always indicate that the model fits the data.

Generalized Coefficients of Determination

Cox and Snell (1989), Maddala (1983), and Magee (1990) proposed a generalization of the coefficient of determination to a general linear model:

$$R^2 = 1 - \left[\frac{L(0)}{L(\hat{\beta})}\right]^{\frac{2}{n}}$$

where L(0) is the likelihood of the intercepts-only model, $L(\hat{\beta})$ is the likelihood of the specified model, and *n* is the sample size. This measure achieves a maximum of less than 1 for discrete models, with maximum given by

$$R^2_{max} = 1 - [L(0)]^{\frac{2}{n}}$$

Nagelkerke (1991) proposed an adjusted coefficient, which can achieve a maximum value of 1:

$$R^2_{adj} = \frac{R^2}{R^2_{max}}$$

Properties and interpretation of R^2 and R^2_{qdj} are provided in Nagelkerke (1991).

The RSQ option, which is available in Release 6.10 and later releases, computes generalized coefficients of determination.

Program

Create the DIABETES data set. The complete DIABETES data set is in the Introduction.

```
data diabetes;
   input patient relwt glufast glutest instest sspg group;
   label relwt   = 'Relative weight'
         glufast = 'Fasting Plasma Glucose'
         glutest = 'Test Plasma Glucose'
         instest = 'Plasma Insulin during Test'
         sspg    = 'Steady State Plasma Glucose'
         group   = 'Clinical Group';
   datalines;
1  0.81  80  356 124   55  1
2  0.95  97  289 117   76  1
3  0.94 105  319 143  105  1
more data lines
;
```

Convert the ordinal response (GROUP) to a binary response (GRP). Combine the chemical diabetics and overt diabetics into one group — the event group. The normals are the nonevent group.

```
data diabet2;
   set diabetes;
   grp=(group=1);
run;
```

Print out the Hosmer and Lemeshow goodness-of-fit test for a model with two explanatory variables. Use the LACKFIT option in the MODEL statement. Specify RSQ to print out the generalized R-square and adjusted R-square values for this model.

```
proc logistic data=diabet2;
   model grp=instest sspg / lackfit rsq;
   title 'Diabetes Data';
run;
```

Print out the Hosmer and Lemeshow goodness-of-fit test and the R-square values for a model with just one explanatory variable.

```
proc logistic data=diabet2;
   model grp=instest / lackfit rsq;
run;
```

Output

Output 7.1
Hosmer and Lemeshow Goodness-of-Fit Test for Two-Variable Model

Diabetes Data

The LOGISTIC Procedure

Data Set: WORK.DIABET2
Response Variable: GRP
Response Levels: 2
Number of Observations: 145
Link Function: Logit

Response Profile

Ordered Value	GRP	Count
1	0	69
2	1	76

Testing Global Null Hypothesis: BETA=0

Criterion	Intercept Only	Intercept and Covariates	Chi-Square for Covariates
AIC	202.675	112.543	.
SC	205.651	121.474	.
-2 LOG L	200.675	106.543	94.131 with 2 DF (p=0.0001)
Score	.	.	72.321 with 2 DF (p=0.0001)

❶ RSquare = 0.4775 ❷ Adjusted RSquare = 0.6372

Analysis of Maximum Likelihood Estimates

Variable	DF	Parameter Estimate	Standard Error	Wald Chi-Square	Pr > Chi-Square	Standardized Estimate	Odds Ratio
INTERCPT	1	-4.6491	0.7642	37.0086	0.0001	.	.
INSTEST	1	0.000799	0.00211	0.1429	0.7054	0.053294	1.001
SSPG	1	0.0249	0.00401	38.6773	0.0001	1.456817	1.025

Association of Predicted Probabilities and Observed Responses

Concordant	= 91.6%	Somers' D	= 0.832
Discordant	= 8.3%	Gamma	= 0.833
Tied	= 0.1%	Tau-a	= 0.418
(5244 pairs)		c	= 0.916

❸ Hosmer and Lemeshow Goodness-of-Fit Test

		GRP = 0		GRP = 1	
Group	Total	Observed	Expected	Observed	Expected
1	15	1	0.52	14	14.48
2	15	0	1.02	15	13.98
3	15	1	1.80	14	13.20
4	15	3	2.93	12	12.07
5	15	7	4.86	8	10.14
6	15	9	8.10	6	6.90
7	15	11	11.89	4	3.11
8	15	12	13.47	3	1.53
9	15	15	14.46	0	0.54
10	10	10	9.95	0	0.05

Goodness-of-fit Statistic = 6.058 with 8 DF (p=0.6407)

Output 7.2
Hosmer and Lemeshow Goodness-of-Fit Test for One-Variable Model

```
                              Diabetes Data

                          The LOGISTIC Procedure

    Data Set: WORK.DIABET2
    Response Variable: GRP
    Response Levels: 2
    Number of Observations: 145
    Link Function: Logit

                             Response Profile

                          Ordered
                            Value     GRP     Count

                                1       0        69
                                2       1        76


                  Testing Global Null Hypothesis: BETA=0

                                 Intercept
                    Intercept       and
   Criterion          Only       Covariates    Chi-Square for Covariates

   AIC              202.675       202.658        .
   SC               205.651       208.611        .
   -2 LOG L         200.675       198.658       2.017 with 1 DF (p=0.1555)
   Score               .             .          1.996 with 1 DF (p=0.1577)

         ❹  RSquare = 0.0138      ❺  Adjusted RSquare = 0.0184

                Analysis of Maximum Likelihood Estimates

            Parameter Standard    Wald       Pr >    Standardized     Odds
  Variable DF  Estimate   Error  Chi-Square Chi-Square   Estimate      Ratio

  INTERCPT 1   -0.4678   0.3138     2.2221     0.1360          .         .
  INSTEST  1   0.00200  0.00144     1.9298     0.1648     0.133200    1.002

       Association of Predicted Probabilities and Observed Responses

                   Concordant = 48.7%          Somers' D = -.014
                   Discordant = 50.1%          Gamma     = -.014
                   Tied       =  1.2%          Tau-a     = -.007
                   (5244 pairs)                c         = 0.493

                  Hosmer and Lemeshow Goodness-of-Fit Test

                                   GRP = 0               GRP = 1
                            --------------------  --------------------
       Group       Total    Observed    Expected  Observed    Expected

           1          16          13        6.49         3        9.51
           2          15           9        6.45         6        8.55
           3          16           5        7.10        11        8.90
           4          16           5        7.22        11        8.78
           5          15           6        6.89         9        8.11
           6          15           3        7.09        12        7.91
           7          16           5        7.80        11        8.20
           8          15           5        7.63        10        7.37
           9          21          18       12.33         3        8.67

      ❻  Goodness-of-fit Statistic = 29.936 with 7 DF (p=0.0001)
```

Explanation

Output 7.1 shows the results from the two-variable binary logistic regression model for the revised diabetes data. At the bottom of the **`Testing Global Null Hypothesis: BETA=0`** table, the two generalized coefficients of determination are printed. The generalized coefficient of determination ❶ has a value of 0.4775. The adjusted generalized coefficient of determination ❷ has a value of 0.6372.

The Hosmer and Lemeshow goodness-of-fit test is printed at the bottom of the output ❸. For this model, the data are divided into nine groups of 15 observations each, and one final group of 10 observations, for a total of 10 groups. The test statistic has a value of 6.058 with eight degrees of freedom. When this statistic is compared to a chi-square distribution, the resulting probability value is 0.6407. Thus, you cannot reject the null hypothesis that the model provides a good fit to the data. However, note that about 50% of the expected frequencies are less than five. For these data, the adequacy of the chi-square distribution is questionable.

Output 7.2 shows the results from using just one explanatory variable to perform logistic regression on the revised diabetes data. The generalized coefficient of determination ❹ and the adjusted coefficient ❺ have values of 0.0138 and 0.0184, respectively. The Hosmer and Lemeshow goodness-of-fit test statistic ❻ has a value of 29.936 with 7 degrees of freedom. This statistic has a probability value of 0.0001. Given this probability value, you can reject the null hypothesis that the one-variable model fits the data well.

References

- Cox, D.R. and Snell, E.J. (1989), *The Analysis of Binary Data: Second Edition*, London: Chapman and Hall.
- Hosmer, D.W. and Lemeshow, S. (1989), *Applied Logistic Regression*, New York: John Wiley & Sons, Inc.
- Maddala, G.S. (1983), *Limited-Dependent and Qualitative Variables in Econometrics*, London: Cambridge University Press.
- Magee, L. (1990), "R^2 Measures Based on Wald and Likelihood Ratio Joint Significance Tests," *American Statistician*, 44, 250-253.
- Nagelkerke, N.J.D. (1991), "A Note on a General Definition of the Coefficient of Determination," *Biometrika*, 78, 691-692.

EXAMPLE 8

Producing Regression Diagnostics for Logistic Regression Models

Featured Tools:
PROC LOGISTIC, MODEL statement:

- ☐ INFLUENCE option
- ☐ IPLOTS option

Regression diagnostics tell you how influential each observation is to the fit of the logistic regression model. Pregibon (1981) developed several measures of influence:

Pearson and Deviance Residuals
: identify observations that are not well explained by the model. Pearson residuals are components of the Pearson chi-square statistic; the Pearson chi-square statistic is the sum of squares of the Pearson residuals. Deviance residuals are components of the deviance, which is another goodness-of-fit statistic based on the log likelihood function.

Hat Matrix Diagonal
: detects extreme points in the design space where they tend to have larger values.

Dfbeta
: assesses the effect of an individual observation on the estimated parameter of the fitted model. A Dfbeta diagnostic is computed for each observation for each parameter estimate. It is the standardized difference in the parameter estimate due to deleting the corresponding observation. The Dfbetas are useful in detecting observations that cause instability in the selected coefficients.

C and CBAR
: provide scalar measures of the influence of individual observations on the parameter estimates. These confidence interval displacement diagnostics are based on the same idea as the Cook distance in linear regression diagnostics.

DIFDEV and DIFCHISQ
: detect ill-fitted observations, that is, observations that contribute heavily to the disagreement between the data and the predicted values of the fitted model. DIFDEV is the change in the deviance due to deleting an individual observation. DIFCHISQ is the change in the Pearson chi-square statistic for the same deletion.

This example uses options in the MODEL statement of the LOGISTIC procedure to compute influence diagnostics for a logistic regression model.

Program

Create the ESR data set. The complete ESR data set is in the Introduction.

```
data esr;
   input id fibrin globulin response @@;
   datalines;
1  2.52 38 0   2  2.56 31 0   3  2.19 33 0   4  2.18 31 0
more data lines
;
```

Print out regression diagnostics. Use the INFLUENCE and IPLOTS options in the MODEL statement.

```
proc logistic data=esr descending;
   model response=fibrin / influence iplots;
   title 'ESR Data';
run;
```

Output

Output 8.1
Regression Diagnostics for Logistic Regression

```
                                 ESR Data

                          The LOGISTIC Procedure

              Data Set: WORK.ESR
              Response Variable: RESPONSE
              Response Levels: 2
              Number of Observations: 32
              Link Function: Logit


                             Response Profile

                        Ordered
                          Value  RESPONSE     Count

                            1         1          6
                            2         0         26

                                 ESR Data

                          The LOGISTIC Procedure

                   Testing Global Null Hypothesis: BETA=0

                                    Intercept
                         Intercept     and
             Criterion     Only     Covariates    Chi-Square for Covariates

             AIC          32.885      28.840          .
             SC           34.351      31.772          .
             -2 LOG L     30.885      24.840         6.045 with 1 DF (p=0.0139)
             Score          .           .            6.752 with 1 DF (p=0.0094)

                    Analysis of Maximum Likelihood Estimates

                   Parameter  Standard     Wald        Pr >      Standardized      Odds
 Variable    DF     Estimate     Error  Chi-Square  Chi-Square     Estimate       Ratio

 INTERCPT    1       -6.8451    2.7703      6.1053      0.0135            .           .
 FIBRIN      1        1.8271    0.9009      4.1134      0.0425     0.641734       6.216

          Association of Predicted Probabilities and Observed Responses

                    Concordant = 71.2%          Somers' D = 0.429
                    Discordant = 28.2%          Gamma     = 0.432
                    Tied       =  0.6%          Tau-a     = 0.135
                    (156 pairs)                 c         = 0.715
```

```
                                                Regression Diagnostics

                              Pearson Residual                 Deviance Residual                 Hat Matrix Diagonal
           Covariates
 Case                           (1 unit = 0.57)                   (1 unit = 0.31)                    (1 unit = 0.02)
Number     FIBRIN      Value  -8  -4  0 2 4 6 8       Value  -8  -4  0 2 4 6 8       Value    0 2 4 6 8  12  16

   1       2.5200    -0.3261  |      *|        |    -0.4496  |      *|        |     0.0406  | *                  |
   2       2.5600    -0.3383  |      *|        |    -0.4655  |     * |        |     0.0402  | *                  |
   3       2.1900    -0.2413  |       *        |    -0.3364  |      *|        |     0.0429  | *                  |
   4       2.1800    -0.2391  |       *        |    -0.3334  |      *|        |     0.0429  | *                  |
   5       3.4100    -0.7354  |      *|        |    -0.9298  |    *  |        |     0.0876  |      *             |
   6       2.4600    -0.3087  |      *|        |    -0.4267  |      *|        |     0.0412  | *                  |
   7       3.2200    -0.6182  |      *|        |    -0.8046  |    *  |        |     0.0589  |  *                 |
   8       2.2100    -0.2457  |       *        |    -0.3424  |      *|        |     0.0428  | *                  |
   9       3.1500    -0.5799  |      *|        |    -0.7614  |     * |        |     0.0520  |  *                 |
  10       2.6000    -0.3509  |      *|        |    -0.4819  |     * |        |     0.0397  | *                  |
  11       2.2900    -0.2643  |       *        |    -0.3675  |      *|        |     0.0426  | *                  |
  12       2.3500    -0.2792  |       *        |    -0.3875  |      *|        |     0.0422  | *                  |
  13       5.0600     0.3012  |       |*       |     0.4167  |       |*       |     0.2770  |                   *|
  14       3.3400     1.4496  |       |  *     |     1.5046  |       |    *   |     0.0753  |   *                |
  15       2.3800     3.4845  |       |     *  |     2.2697  |       |      * |     0.0420  | *                  |
  16       3.1500    -0.5799  |      *|        |    -0.7614  |     * |        |     0.0520  |  *                 |
  17       3.5300     1.2186  |       | *      |     1.3493  |       |   *    |     0.1135  |        *           |
  18       2.6800    -0.3775  |      *|        |    -0.5162  |     * |        |     0.0390  | *                  |
  19       2.6000    -0.3509  |      *|        |    -0.4819  |     * |        |     0.0397  | *                  |
  20       2.2300    -0.2502  |       *        |    -0.3485  |      *|        |     0.0428  | *                  |
  21       2.8800    -0.4531  |      *|        |    -0.6112  |     * |        |     0.0394  | *                  |
  22       2.6500    -0.3673  |      *|        |    -0.5031  |     * |        |     0.0393  | *                  |
  23       2.0900     4.5414  |       |       *|     2.4794  |       |       *|     0.0427  | *                  |
  24       2.2800    -0.2619  |       *        |    -0.3643  |      *|        |     0.0426  | *                  |
  25       2.6700    -0.3740  |      *|        |    -0.5118  |     * |        |     0.0391  | *                  |
  26       2.2900    -0.2643  |       *        |    -0.3675  |      *|        |     0.0426  | *                  |
  27       2.1500    -0.2326  |       *        |    -0.3246  |      *|        |     0.0428  | *                  |
  28       2.5400    -0.3322  |      *|        |    -0.4575  |      *|        |     0.0404  | *                  |
  29       3.9300     0.8456  |       |*       |     1.0387  |       |  *     |     0.2265  |                 *  |
  30       3.3400    -0.6898  |      *|        |    -0.8823  |     * |        |     0.0753  |   *                |
  31       2.9900    -0.5010  |      *|        |    -0.6693  |     * |        |     0.0423  | *                  |
  32       3.3200    -0.6773  |      *|        |    -0.8690  |     * |        |     0.0721  |   *                |
```

```
                                       Regression Diagnostics

                  INTERCPT Dfbeta                  FIBRIN  Dfbeta                       C

 Case                (1 unit = 0.11)                 (1 unit =  0.1)                     (1 unit = 0.06)
Number     Value   -8  -4  0 2 4 6 8      Value   -8  -4  0 2 4 6 8      Value       0 2 4 6 8  12  16

   1     -0.0540   |       *        |    0.0453   |       *        |    0.00469    |*                  |
   2     -0.0539   |       *        |    0.0445   |       *        |    0.00499    |*                  |
   3     -0.0478   |       *        |    0.0431   |       *        |    0.00272    |*                  |
   4     -0.0475   |       *        |    0.0428   |       *        |    0.00267    |*                  |
   5      0.0956   |       |*       |   -0.1344   |      *|        |     0.0569    | *                 |
   6     -0.0538   |       *        |    0.0460   |       *        |    0.00428    |*                  |
   7      0.0236   |       *        |   -0.0524   |      *|        |     0.0254    |*                  |
   8     -0.0484   |       *        |    0.0435   |       *        |    0.00282    |*                  |
   9     0.00489   |       *        |   -0.0306   |       *        |     0.0194    |*                  |
  10     -0.0535   |       *        |    0.0434   |       *        |    0.00531    |*                  |
  11     -0.0507   |       *        |    0.0450   |       *        |    0.00325    |*                  |
  12     -0.0521   |       *        |    0.0457   |       *        |    0.00359    |*                  |
  13     -0.1967   |     * |        |    0.2112   |       | *      |     0.0481    | *                 |
  14     -0.1368   |      *|        |    0.2098   |       | *      |     0.1850    |   *               |
  15      0.6396   |       |     *  |   -0.5578   |  *    |        |     0.5554    |        *          |
  16     0.00489   |       *        |   -0.0306   |       *        |     0.0194    |*                  |
  17     -0.2392   |     * |        |    0.3084   |       |  *     |     0.2145    |   *               |
  18     -0.0516   |       *        |    0.0400   |       *        |    0.00602    |*                  |
  19     -0.0535   |       *        |    0.0434   |       *        |    0.00531    |*                  |
  20     -0.0490   |       *        |    0.0439   |       *        |    0.00293    |*                  |
  21     -0.0391   |       *        |    0.0227   |       *        |    0.00878    |*                  |
  22     -0.0525   |       *        |    0.0415   |       *        |    0.00574    |*                  |
  23      0.9166   |       |       *|   -0.8366   |*      |        |     0.9605    |                  *|
  24     -0.0504   |       *        |    0.0448   |       *        |    0.00319    |*                  |
  25     -0.0519   |       *        |    0.0405   |       *        |    0.00593    |*                  |
```

```
   26   -0.0507  |        *        |   0.0450  |       *        |   0.00325  |*               |
   27   -0.0465  |        *        |   0.0421  |       *        |   0.00253  |*               |
   28   -0.0540  |        *        |   0.0450  |       *        |   0.00484  |*               |
   29   -0.3821  |     *  |        |   0.4408  |       |   *    |    0.2706  |      *         |
   30    0.0651  |        |*       |  -0.0999  |      *|        |    0.0419  | *              |
   31   -0.0258  |        *        |  0.00606  |       *        |    0.0116  |*               |
   32    0.0573  |        *        |  -0.0910  |      *|        |    0.0384  | *              |
```

```
                                     Regression Diagnostics

                         CBAR                          DIFDEV                          DIFCHISQ

  Case                (1 unit = 0.06)             (1 unit = 0.44)               (1 unit = 1.35)
 Number      Value    0 2 4 6 8  12  16     Value  0 2 4 6 8  12  16     Value   0 2 4 6 8  12  16

     1     0.00450   |*               |   0.2067  |*               |   0.1109   |*               |
     2     0.00479   |*               |   0.2215  | *              |   0.1192   |*               |
     3     0.00261   |*               |   0.1158  |*               |   0.0608   |*               |
     4     0.00256   |*               |   0.1137  |*               |   0.0597   |*               |
     5      0.0519   | *              |   0.9165  |  *             |   0.5927   |*               |
     6     0.00410   |*               |   0.1862  |*               |   0.0994   |*               |
     7      0.0239   |*               |   0.6712  |  *             |   0.4061   |*               |
     8     0.00270   |*               |   0.1199  |*               |   0.0631   |*               |
     9      0.0184   |*               |   0.5982  | *              |   0.3547   |*               |
    10     0.00510   |*               |   0.2373  | *              |   0.1282   |*               |
    11     0.00311   |*               |   0.1382  |*               |   0.0730   |*               |
    12     0.00344   |*               |   0.1536  |*               |   0.0814   |*               |
    13      0.0348   | *              |   0.2084  |*               |   0.1255   |*               |
    14      0.1711   |   *            |   2.4348  |      *         |   2.2725   |  *             |
    15      0.5321   |        *       |   5.6836  |            *   |  12.6735   |        *       |
    16      0.0184   |*               |   0.5982  | *              |   0.3547   |*               |
    17      0.1901   |   *            |   2.0107  |     *          |   1.6752   | *              |
    18     0.00579   |*               |   0.2722  | *              |   0.1483   |*               |
    19     0.00510   |*               |   0.2373  | *              |   0.1282   |*               |
    20     0.00280   |*               |   0.1243  |*               |   0.0654   |*               |
    21     0.00843   |*               |   0.3820  | *              |   0.2138   |*               |
    22     0.00551   |*               |   0.2586  | *              |   0.1404   |*               |
    23      0.9195   |               *|   7.0671  |               *|  21.5439   |               *|
    24     0.00306   |*               |   0.1358  |*               |   0.0717   |*               |
    25     0.00569   |*               |   0.2676  | *              |   0.1456   |*               |
    26     0.00311   |*               |   0.1382  |*               |   0.0730   |*               |
    27     0.00242   |*               |   0.1078  |*               |   0.0565   |*               |
    28     0.00464   |*               |   0.2140  |*               |   0.1150   |*               |
    29      0.2093   |    *           |   1.2883  |   *            |   0.9244   | *              |
    30      0.0387   | *              |   0.8172  |  *             |   0.5146   |*               |
    31      0.0111   |*               |   0.4590  | *              |   0.2621   |*               |
    32      0.0357   | *              |   0.7909  |  *             |   0.4945   |*               |
```

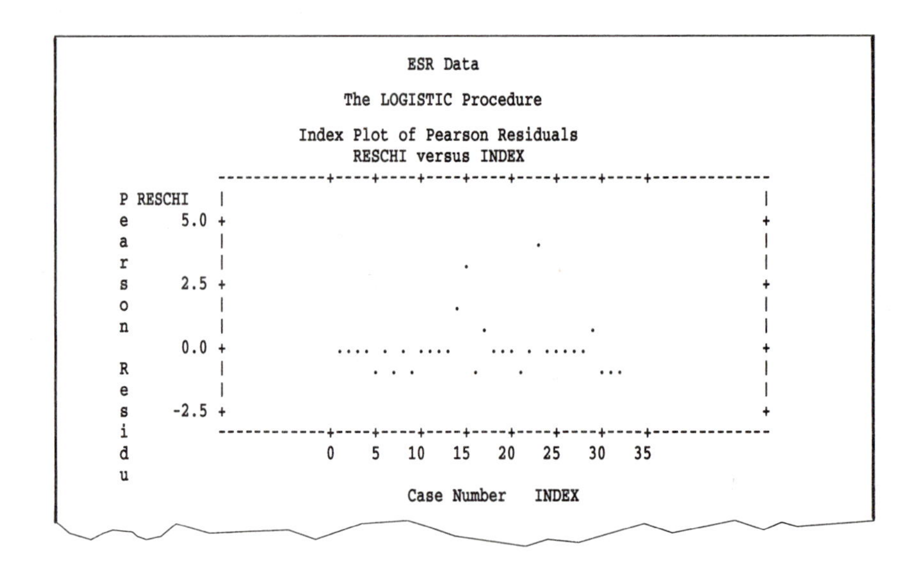

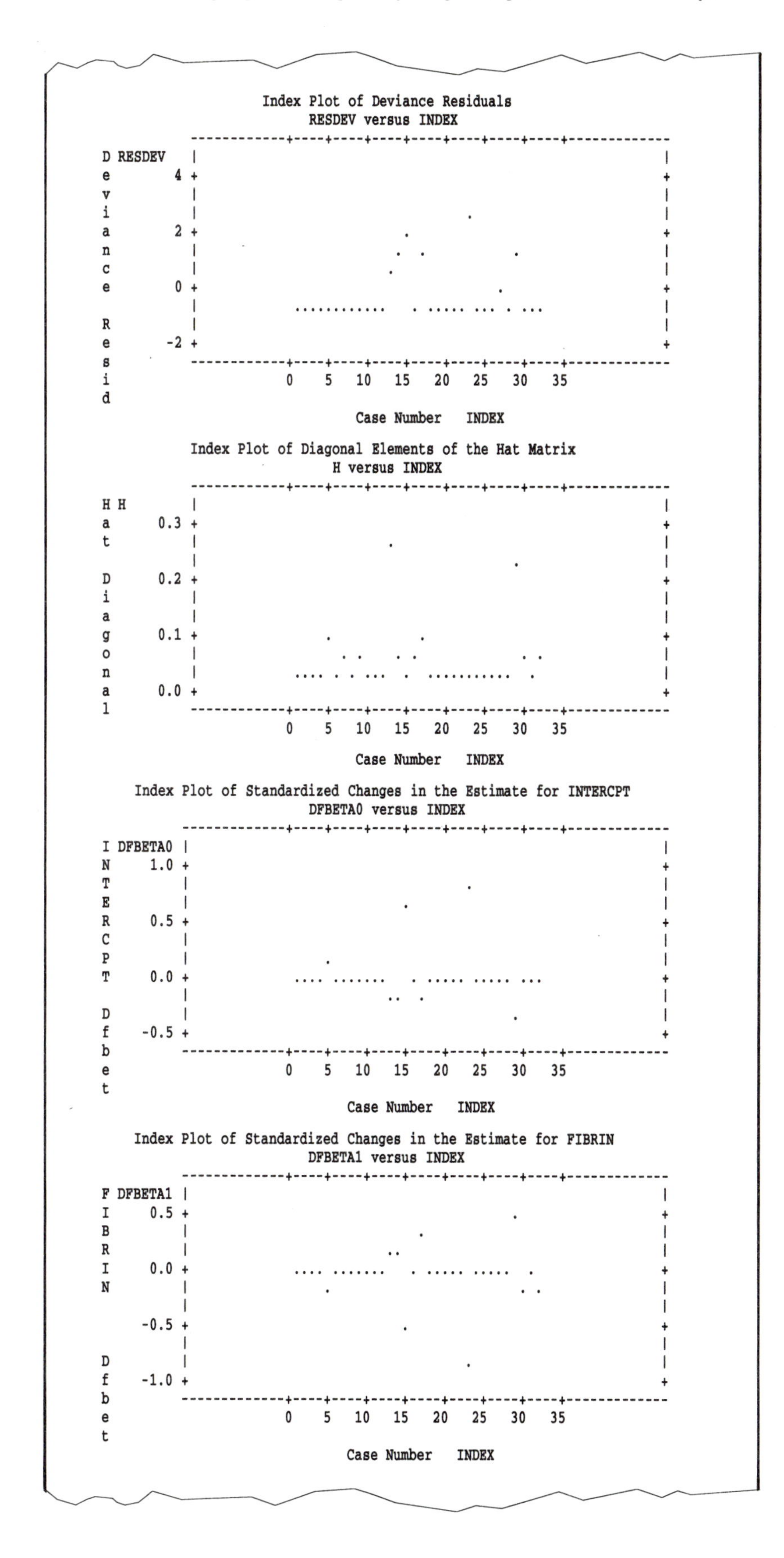
Index Plot of Deviance Residuals
RESDEV versus INDEX
Deviance Resid
RESDEV
4
2
0
-2
0 5 10 15 20 25 30 35
Case Number INDEX
Index Plot of Diagonal Elements of the Hat Matrix
H versus INDEX
Hat Diagonal
H
0.3
0.2
0.1
0.0
0 5 10 15 20 25 30 35
Case Number INDEX
Index Plot of Standardized Changes in the Estimate for INTERCPT
DFBETA0 versus INDEX
INTERCPT Dfbet
DFBETA0
1.0
0.5
0.0
-0.5
0 5 10 15 20 25 30 35
Case Number INDEX
Index Plot of Standardized Changes in the Estimate for FIBRIN
DFBETA1 versus INDEX
FIBRIN Dfbet
DFBETA1
0.5
0.0
-0.5
-1.0
0 5 10 15 20 25 30 35
Case Number INDEX

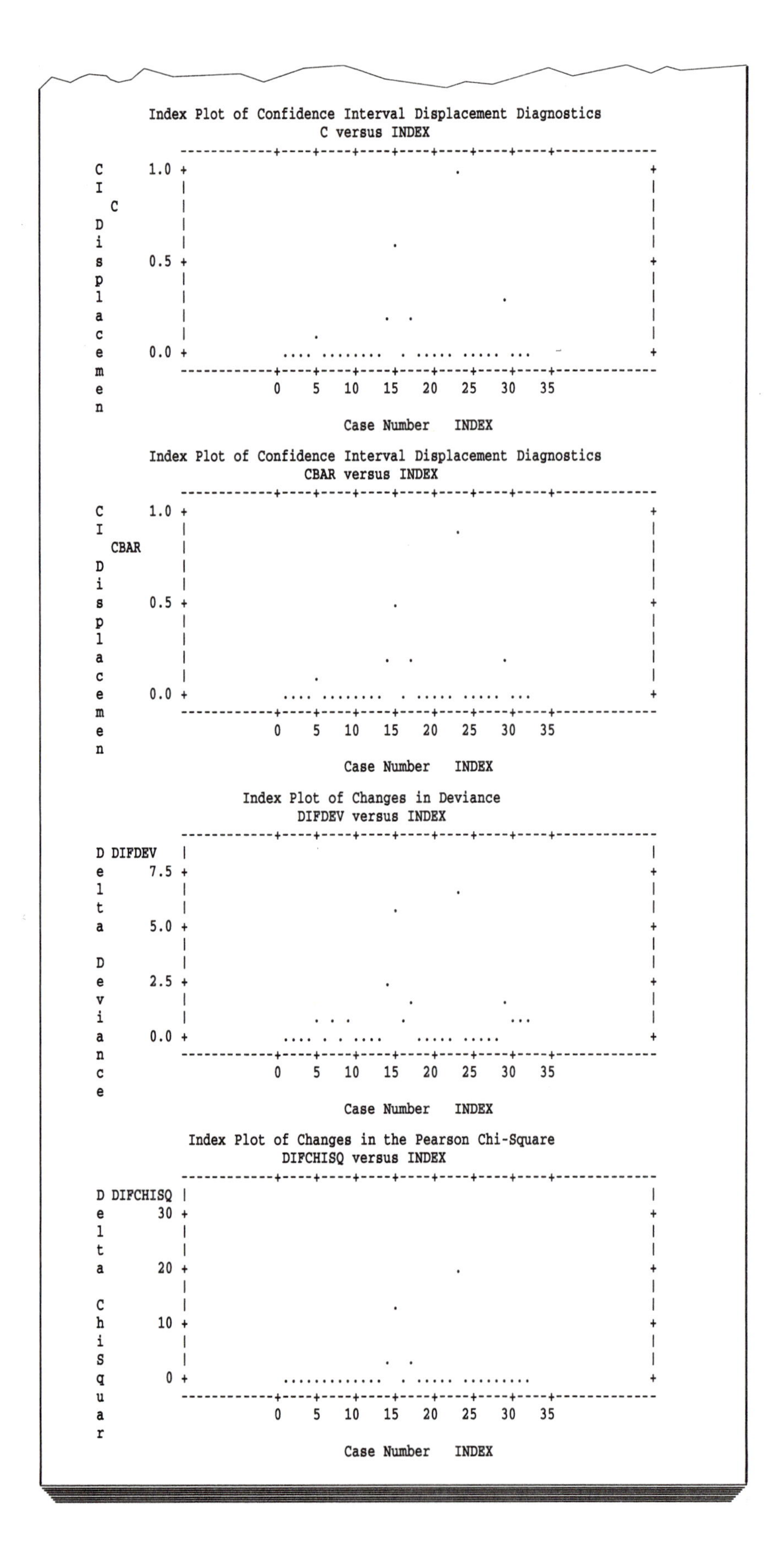

Index Plot of Confidence Interval Displacement Diagnostics
C versus INDEX
CI C Displacement
1.0
0.5
0.0
0 5 10 15 20 25 30 35
Case Number INDEX
Index Plot of Confidence Interval Displacement Diagnostics
CBAR versus INDEX
CI CBAR Displacement
1.0
0.5
0.0
0 5 10 15 20 25 30 35
Case Number INDEX
Index Plot of Changes in Deviance
DIFDEV versus INDEX
Delta Deviance DIFDEV
7.5
5.0
2.5
0.0
0 5 10 15 20 25 30 35
Case Number INDEX
Index Plot of Changes in the Pearson Chi-Square
DIFCHISQ versus INDEX
Delta ChiSquar DIFCHISQ
30
20
10
0
0 5 10 15 20 25 30 35
Case Number INDEX

Explanation

When you specify the INFLUENCE option in the MODEL statement, PROC LOGISTIC prints the case number (observation number), values of all explanatory variables included in the final model, and all of Pregibon's regression diagnostic measures. The IPLOTS option in the MODEL statement causes PROC LOGISTIC to print index plots of each regression diagnostic measure against each case number.

From the listings and plots shown in Output 8.1, you can see that some observations are very influential in fitting the model. For example, observations 15 and 23 have large values for most of the diagnostics, indicating that these observations are very influential to the fit of the model. It may be worthwhile to examine these observations for any possible transcription errors. In this case, observations 15 and 23, along with observations 13, 14, 17, and 29, are the only observations with positive values for the response variable. That is, these six observations represent the event, and the other 26 observations represent the nonevent. A small number of event observations can often exert a strong influence on the fit of the model. See Collett (1991) for a complete discussion of logistic regression diagnostics.

References

□ Collett, D.R. (1991), *Modeling Binary Data*, London: Chapman and Hall.

□ Pregibon, D. (1981), "Logistic Regression Diagnostics," *Annals of Statistics*, 9, 705-724.

E X A M P L E 9

Correcting for Overdispersion in Logistic Regression Models

Featured Tools:
PROC LOGISTIC, MODEL statement:

- AGGREGATE option
- SCALE= option

Overdispersion occurs when the variance of the response variable exceeds the nominal variance, which is $np(1 - p)$, where n is the sample size and p is the probability of the event for binary outcomes. It is caused by positive correlation between the binary responses, or variation between the response probabilities. Depending on the type of data you collect, the causes of overdispersion can be due to genetic, social, physiological, biochemical, environmental, or other factors. These factors cause the binary responses in your data to become correlated. Treating different experimental groups in different ways also causes overdispersion, leading to variation between the response probabilities. If you could control for these factors and causes, you could reduce or eliminate the overdispersion in your data.

Apparent overdispersion occurs when the systematic component of the model is inadequate in some way, such as when

- you fail to include a sufficient number of interaction terms in a large factorial experiment
- you assume a linear relationship between the logit transform of the response variable and the explanatory variables *and* the actual relationship is quadratic, or higher order
- you should use a log or some other transformation of the explanatory variable
- the data contain outliers
- the model omits important explanatory variables
- the number of observations in each subpopulation is small.

Eliminate these possibilities before you conclude that the data are overdispersed.

Overdispersion, or apparent overdispersion, is common in most real data. It causes you to underestimate the variance of parameter estimates. With overdispersion, the true variance of the response variable is given by $\sigma^2 np(1 - p)$, where σ^2 is the dispersion parameter. Underdispersion is also possible but is less common in real data applications.

To get a correct estimate of the variance, inflate (or deflate) the nominal variance by the dispersion parameter. In some cases, you may know the cause of the overdispersion (or underdispersion). You can compute the dispersion parameter directly and multiply the nominal variance by that amount. In most cases, however, the dispersion parameter is unknown, but you can estimate it.

Two common methods of estimating an unknown dispersion parameter are available. They use the Pearson chi-square statistic χ^2_P and the Deviance chi-square statistic χ^2_D:

- $$\chi^2_P = \sum_{i=1}^{m}\sum_{j=1}^{k} \left(r_{ij} - n_i p_{ij}\right)^2 / n_i p_{ij}$$
- $$\chi^2_D = 2\sum_{i=1}^{m}\sum_{j=1}^{k} r_{ij} \log\left(\frac{r_{ij}}{n_i p_{ij}}\right)$$

where

m	is the number of subpopulation profiles.
k	is the number of response levels.
r_{ij}	is the weight of the response at the *j*th level for the *i*th profile.
n_i	is the total weight at the *i*th profile.
p_{ij}	is the fitted probability for the *j*th level at the *i*th profile.

Each of these statistics approximates a chi-square distribution with $m(k - 1) - q$ degrees of freedom (for large n_i), where q is the number of parameters estimated. The dispersion parameter is estimated by the chi-square divided by the degrees of freedom. If the dispersion parameter is 1, there is no overdispersion or underdispersion. A dispersion parameter greater than 1 indicates overdispersion; a value less than 1 indicates underdispersion. In the case of $n_i = 1$, the Pearson and Deviance statistics do not approximate a chi-square distribution. In that case, you cannot use these statistics to test for overdispersion.

This example uses the AGGREGATE and SCALE= options in the MODEL statement of the LOGISTIC procedure to correct a logistic regression model for overdispersion. These options are available in Release 6.10 and later releases.

Program

Create the DIABETES data set. The complete DIABETES data set is in the Introduction.

```
data diabetes;
   input patient relwt glufast glutest instest sspg group;
   label relwt   = 'Relative weight'
         glufast = 'Fasting Plasma Glucose'
         glutest = 'Test Plasma Glucose'
         instest = 'Plasma Insulin during Test'
         sspg    = 'Steady State Plasma Glucose'
         group   = 'Clinical Group';
   datalines;
1  0.81  80  356 124   55  1
2  0.95  97  289 117   76  1
3  0.94 105  319 143  105  1
more data lines
;
```

Convert the ordinal response (GROUP) to a binary response (GRP). Combine the chemical diabetics and overt diabetics into one group — the event group. The normals are the nonevent group.

```
data diabet2;
   set diabetes;
   grp=(group=1);
run;
```

Correct the logistic regression model for overdispersion. Use the AGGREGATE and SCALE= options in the MODEL statement.

```
proc logistic data=diabet2;
   model grp=instest / aggregate scale=deviance;
   title 'Diabetes Data';
run;
```

Compare the results to a model uncorrected for overdispersion.

```
proc logistic data=diabet2;
   model grp=instest;
run;
```

Output

Output 9.1
Correcting for Overdispersion in Logistic Regression

```
                          Diabetes Data

                      The LOGISTIC Procedure

        Data Set: WORK.DIABET2
        Response Variable: GRP
        Response Levels: 2
        Number of Observations: 145
        Link Function: Logit


                         Response Profile

                     Ordered
                      Value      GRP      Count

                        1          0        69
                        2          1        76

  ❶  Deviance and Pearson Goodness-of-Fit Statistics

                                                          Pr >
  Criterion        DF        Value     Value/DF     Chi-Square
                                          ❷
  Deviance        112        160.5       1.4332         0.0018
  Pearson         112        117.5       1.0495         0.3414
```

```
                    Number of unique profiles: 114

❸ NOTE: The covariance matrix has been multiplied by the heterogeneity factor
     1.43322.

                 Testing Global Null Hypothesis: BETA=0

                                Intercept
                   Intercept       and
 Criterion           Only       Covariates    Chi-Square for Covariates

 AIC               142.016       142.609           .
 SC                144.993       148.562           .
 -2 LOG L          140.016       138.609        1.407 with 1 DF (p=0.2355)
 Score                .             .           1.393 with 1 DF (p=0.2380)

                 Analysis of Maximum Likelihood Estimates

              Parameter Standard     Wald       Pr >    Standardized     Odds
 Variable DF  Estimate    Error  Chi-Square Chi-Square    Estimate       Ratio
                            ❹        ❺          ❻
 INTERCPT 1    -0.4678   0.3757     1.5504     0.2131          .            .
 INSTEST  1    0.00200  0.00172     1.3465     0.2459     0.133200      1.002

        Association of Predicted Probabilities and Observed Responses

                 Concordant = 48.7%          Somers' D = -.014
                 Discordant = 50.1%          Gamma     = -.014
                 Tied       =  1.2%          Tau-a     = -.007
                 (5244 pairs)                c         = 0.493
```

Output 9.2
Parameter Estimates and Standard Errors for a Model Uncorrected for Overdispersion (Partial Output)

```
                             Diabetes Data

                         The LOGISTIC Procedure

                 Analysis of Maximum Likelihood Estimates

              Parameter Standard     Wald       Pr >    Standardized     Odds
 Variable DF  Estimate    Error  Chi-Square Chi-Square    Estimate       Ratio

 INTERCPT 1    -0.4678   0.3138     2.2221     0.1360          .            .
 INSTEST  1    0.00200  0.00144     1.9298     0.1648     0.133200      1.002
```

Explanation

In Output 9.1, PROC LOGISTIC prints the **`Deviance and Pearson Goodness-of-Fit Statistics`** table ❶ when you specify the AGGREGATE and SCALE= options. This table lists the name of the criterion, its degrees of freedom, its value, the ratio of the value over the degrees of freedom, and the probability value for the chi-square. The column labeled **`Value/DF`** ❷ contains the estimates of the dispersion parameter. The number of unique subpopulation profiles is listed at the bottom of the table. This number represents the unique combinations of the INSTEST variable, which is 114. A note in the output informs you that the covariance matrix is multiplied by the appropriate dispersion parameter ❸. Because you specify SCALE=DEVIANCE, the covariance matrix is multiplied by the ratio of the Deviance chi-square divided by its degrees of freedom, which is 1.43322.

Note: The Deviance and Pearson statistics fail to approximate a chi-square distribution as the number of subpopulation profiles increases, and consequently, the number of individuals in each subpopulation decreases. Because there are 145 observations in 114 subpopulations, the chi-square approximation is not good for these data.

In the **Analysis of Maximum Likelihood Estimates** table in Output 9.1, the standard errors for the parameter estimates are inflated to account for the overdispersion ❹. This, in turn, affects the Wald Chi-Square values ❺ and the probability values for the Wald Chi-Square ❻. The parameter estimates, standardized estimates, and odds ratios are unaffected by the dispersion parameter. Compare these results to the table shown in Output 9.2 for the model that is uncorrected for overdispersion.

Variations

Using Other Methods to Correct for Overdispersion

If you specify SCALE=PEARSON in the MODEL statement of PROC LOGISTIC, the procedure uses the Pearson chi-square, divided by its degrees of freedom, as the estimated dispersion parameter. Another specification for the SCALE= option is WILLIAMS (available with Release 6.11 and later releases), which can only be used with the events/trials model syntax. Using this method, PROC LOGISTIC estimates the dispersion parameter in an iterative process that involves fitting weighted regression models. See Collett (1991) for more information on the Williams method. You can also specify a constant value in the SCALE= option. In that case, PROC LOGISTIC squares the constant value that you specify and uses that squared value as the estimated dispersion parameter. If you specify SCALE=NONE, PROC LOGISTIC prints the **Deviance and Pearson Goodness-of-Fit Statistics** table but does not adjust the covariance matrix with the estimated dispersion parameter.

You can form subpopulations to compute the Pearson and Deviance chi-square statistics based on the unique combinations of values in one or more variables from the input data set. Use the AGGREGATE= option in the MODEL statement of PROC LOGISTIC to specify the variables to use in forming the subpopulation profiles. The AGGREGATE= option enables you to form subpopulations according to the way the data were collected. For example, suppose you collected data at combinations of levels of three variables: A, B, and C. In an initial analysis you find that C is not an important explanatory variable for your model. You can refit the model excluding C but still use C to define the subpopulations properly by specifying AGGREGATE=(A B C) in the MODEL statement of PROC LOGISTIC. Specifying AGGREGATE (without the equal sign and the variable list) is equivalent to specifying AGGREGATE= with a variable list that includes all explanatory variables in the MODEL statement.

PROC LOGISTIC ignores the SCALE= option unless you define subpopulations either with the AGGREGATE= option or by using events/trials syntax in the MODEL statement.

Reference

□ Collett, D.R. (1991), *Modeling Binary Data*, London: Chapman and Hall.

EXAMPLE 10

Displaying an ROC Curve

Featured Tools:

- PROC LOGISTIC, MODEL statement:
  events/trials syntax
  OUTROC= option
- GPLOT procedure

ROC Curves

The receiver operating characteristic (ROC) curve is a graphic display that gives a measure of the predictive accuracy of a logistic regression model. For a logistic regression model with high predictive accuracy, the ROC curve rises quickly. Thus, the area under the curve is large for a model with high predictive accuracy. Conversely, the ROC curve rises slowly and has a smaller area under the curve for logistic regression models with low predictive accuracy.

The area under the ROC curve is not an extremely sensitive measure to use when you compare models. For best results, compute the area under the ROC curve with independent data that were not used to fit the competing models. You can also use bootstrapping or crossvalidation methods to correct for the bias that occurs when you use the same data to test the accuracy of the model that you use to fit the model.

To understand how the ROC curve is plotted, you should understand two measures of predictive accuracy and one count of predictive inaccuracy:

- *Sensitivity* is a measure of accuracy for predicting events. It is the proportion of event observations that the model predicts to be events for a given probability cutpoint.
- *Specificity* is a measure of accuracy for predicting nonevents. It is the proportion of nonevent observations that the model predicts to be nonevents for a given probability cutpoint.
- *False positives* are the number of nonevent observations that the model incorrectly predicts as events for a given probability cutpoint. The number of correctly predicted nonevents added to the number of false positives gives the total number of nonevent observations in the sample.

Sensitivity and specificity vary according to the probability value you choose as a cutpoint for deciding if an observation represents an event or a nonevent. For example, if you use a rule that all observations with predicted probabilities of 0 or higher be classified as events (a cutpoint of 0), then all event observations are correctly classified as events. (In fact, all of the observations are classified as events.) That is, the sensitivity is 1. However, specificity is 0, because none of the nonevent observations are correctly classified as nonevents. If you use a rule that all observations with predicted probabilities above 1 be classified as events (a cutpoint of 1), then none of the event observations are correctly classified as events, but all of the nonevent observations are correctly classified as nonevents. That is, the sensitivity is 0 and the specificity is 1. As you increase the probability cutpoint from 0 to 1, the sensitivity decreases and the specificity increases. Ideally, you would like to have high values for both sensitivity and specificity, so that your model can accurately predict both events and nonevents. Often, the rates of change in the sensitivity and specificity are different, so that for some probability cutpoints, both sensitivity and specificity can have high values. The estimated event probabilities used to compute the sensitivity and specificity for plotting ROC curves are not adjusted for bias.

See Figure 10.1 for an example of an ROC curve.

Figure 10.1
Sample ROC Curve

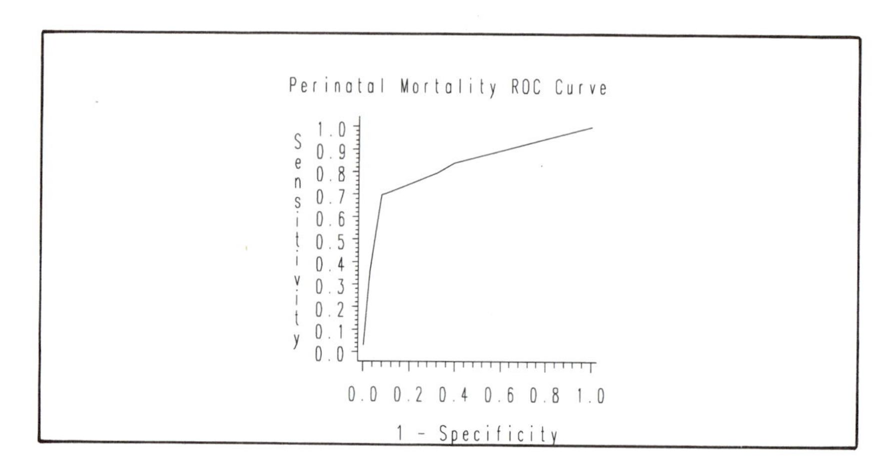

The ROC curve is a plot of the sensitivity against 1 minus the specificity. The data set used to display the ROC curve lists the estimated event probabilities in descending order so that both sensitivity and 1 minus specificity increase as the probabilities decrease. Another way of describing 1 minus the specificity is that it is the number of false positives divided by the number of nonevents. The ROC curve rises quickly when both sensitivity and specificity are high for the higher estimated probability cutpoint values. You create the output data set for displaying the ROC curve with the OUTROC= option, which is available in Release 6.10 and later releases.

See Figure 10.2 for an example of an ROC curve that does not rise quickly. The model used to create the ROC curve in Figure 10.2 has low predictive accuracy. (It uses the diabetes data with INSTEST as the only explanatory variable.) The *c* statistic for this ROC curve is 0.493. See Example 7 and Example 9 for more information on the model used to create the ROC curve shown in Figure 10.2.

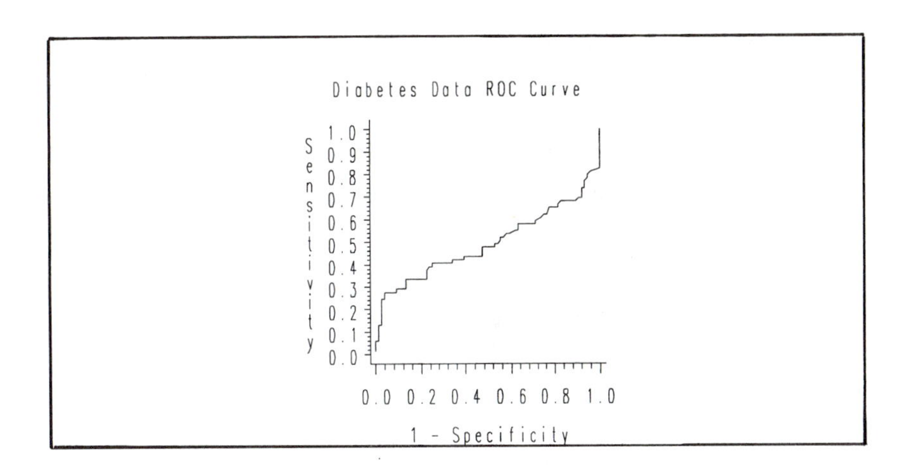

Figure 10.2
A Slowly Rising ROC Curve

Events/Trials Syntax

This example uses the events/trials syntax in the MODEL statement of PROC LOGISTIC. In the events/trials model syntax, which can only be used for binary response data, you specify two variables that contain count data for a binomial experiment. These two variables are separated by a slash. The value of the first variable, *events*, is the number of positive responses (or events). The value of the second variable, *trials*, is the number of trials. The values of both *events* and *trials - events* must be nonnegative, and the value of *trials* must be positive for the response to be valid.

Program

Create the MORTAL data set. The complete MORTAL data set is in the Introduction.

```
data mortal;
   input deaths tbirths cigs age gestpd;
   datalines;
50 365  1 1 1
9  49   2 1 1
41 188  1 2 1
more data lines
;
```

Create an output data set that contains the data necessary to display the ROC curve. Use the events/trials syntax and the OUTROC= option in the MODEL statement.

```
proc logistic data=mortal;
   model deaths/tbirths=cigs age gestpd / outroc=roc1;
   title 'Perinatal Mortality Data';
run;
```

Print out the OUTROC= data set.

```
proc print data=roc1;
   title2 'OUTROC= Data Set';
run;
```

Set graphics options.

```
goptions cback=white
         colors=(black)
         border;
```

Specify the axes for the ROC curve.

```
axis1 length=2.5in;
axis2 order=(0 to 1 by .1) length=2.5in;
```

Use PROC GPLOT to display the ROC curve.

```
proc gplot data=roc1;
   symbol1 i=join v=none;
   title1;
   title2 'Perinatal Mortality ROC Curve';
   plot _sensit_*_1mspec_ / haxis=axis1 vaxis=axis2;
run;
```

Output

Output 10.1
Output from PROC LOGISTIC

```
                        Perinatal Mortality Data

                         The LOGISTIC Procedure

   Data Set: WORK.MORTAL
❶ Response Variable (Events): DEATHS
❷ Response Variable (Trials): TBIRTHS
   Number of Observations: 8
   Link Function: Logit


                            Response Profile

                        Ordered  Binary ❸
                          Value  Outcome        Count

                              1  EVENT            149
                              2  NO EVENT        6602

                Testing Global Null Hypothesis: BETA=0

                                   Intercept
                    Intercept         and
   Criterion          Only        Covariates    Chi-Square for Covariates

   AIC              1433.110       1090.833            .
   SC               1439.927       1118.103            .
   -2 LOG L         1431.110       1082.833       348.277 with 3 DF (p=0.0001)
   Score                .              .          683.924 with 3 DF (p=0.0001)

                Analysis of Maximum Likelihood Estimates

              Parameter Standard    Wald        Pr >    Standardized     Odds
   Variable DF Estimate   Error  Chi-Square Chi-Square   Estimate       Ratio

   INTERCPT 1    0.5643   0.4830     1.3648     0.2427          .           .
   CIGS     1    0.4162   0.2621     2.5211     0.1123      0.067883     1.516
   AGE      1    0.4866   0.1805     7.2683     0.0070      0.119385     1.627
   GESTPD   1   -3.2878   0.1847   316.8595     0.0001     -0.522385     0.037

       Association of Predicted Probabilities and Observed Responses

                 Concordant = 75.1%          Somers' D = 0.662
                 Discordant =  8.9%          Gamma     = 0.789
                 Tied       = 16.0%          Tau-a     = 0.029
                 (983698 pairs)            ❹ c         = 0.831
```

Output 10.2
Listing of the OUTROC= Data Set from PROC LOGISTIC

```
                         Perinatal Mortality Data
                            OUTROC= Data Set

OBS    _PROB_    _POS_   _NEG_   _FALPOS_   _FALNEG_   _SENSIT_   _1MSPEC_

 1    0.28540      4     6591       11        145      0.02685    0.00167
 2    0.20849     45     6444      158        104      0.30201    0.02393
 3    0.19711     54     6404      198         95      0.36242    0.02999
 4    0.13935    104     6089      513         45      0.69799    0.07770
 5    0.01469    105     5965      637         44      0.70470    0.09649
 6    0.00974    119     4471     2131         30      0.79866    0.32278
 7    0.00908    125     4012     2590         24      0.83893    0.39231
 8    0.00601    149        0     6602          0      1.00000    1.00000
```

Output 10.3
ROC Curve from Logistic Regression

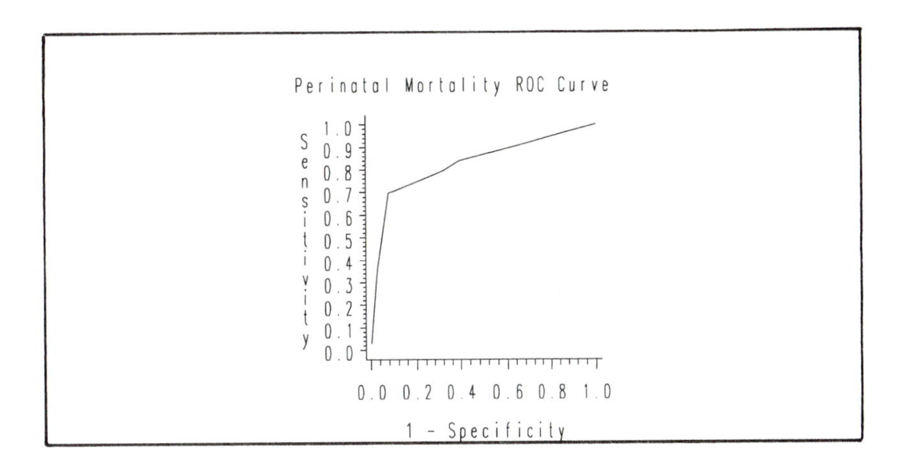

Explanation

Output 10.1 shows the results of the logistic regression analysis for the perinatal mortality data. When you use the events/trials syntax, the beginning of the PROC LOGISTIC output is different than it is when you use the actual model syntax. The output lists two response variables: one for the events variable ❶, and one for the trials variable ❷. In the **`Response Profile`** table, a variable called **`Binary Outcome`** ❸ lists the values **`EVENT`** and **`NO EVENT`**.

The value of the *c* statistic ❹ in the **`Association of Predicted Probabilities and Observed Responses`** table represents the area under the ROC curve (Bamber 1975; Hanley and McNeil 1982). The higher the number, the better the predictive power of the logistic regression model. Given that the two variables in the plot range from 0 to 1, the area under the curve can range from 0 to 1. In this example, the value of the *c* statistic (area) is 0.831.

The OUTROC= data set is listed in Output 10.2. The data set contains:

PROB	the estimated probability of an event. These estimated probabilities serve as cutpoints for predicting the response. Any observation with an estimated event probability that exceeds or equals _PROB_ is predicted to be an event; otherwise it is predicted as a nonevent.
POS	the number of correctly predicted event responses.
NEG	the number of correctly predicted nonevent responses.
FALPOS	the number of falsely predicted event responses.
FALNEG	the number of falsely predicted nonevent responses.
SENSIT	the sensitivity, which is the proportion of event observations that were predicted to have an event response.
1MSPEC	1 minus specificity, which is the proportion of nonevent observations that were predicted to have an event response.

To obtain an ROC curve, plot _SENSIT_ against _1MSPEC_. Output 10.3 shows the ROC curve for the perinatal mortality data. Note that the curve rises fairly quickly, indicating that the predictive accuracy of this logistic regression model is good.

Further Reading

- For complete reference information on the GPLOT procedure, see *SAS/GRAPH Software: Reference, Version 6, First Edition*, Volume 1 and Volume 2.

References

- Bamber, D. (1975), "The Area Above the Ordinal Dominance Graph and the Area Below the Receiver Operating Characteristic Graph," *Journal of Mathematical Psychology*, 12, 387-415.
- Hanley, J.A. and McNeil, B.J. (1982), "The Meaning and Use of the Area Under a Receiver Operating Characteristic (ROC) Curve," *Radiology*, 143, 29-36.

E X A M P L E 11

Infinite Parameters

Featured Tools:

- PROC LOGISTIC
- PROC PLOT
- PROC MEANS

The likelihood equation for a logistic regression model does not always have a finite solution. Statisticians use the term *infinite parameters* to refer to the situation when no finite maximum likelihood estimate exists. The existence, finiteness, and uniqueness of maximum likelihood estimates for the logistic regression model depend upon the patterns of data points in the observation space (Albert and Anderson 1984; Hauck and Donner 1977; Santner and Duffy 1986).

The three mutually exclusive and exhaustive types of data configurations are

Complete Separation
: The maximum likelihood estimate does not exist. There exists a vector of pseudoestimates that correctly allocates all observations to their observed response groups. Such a data configuration gives nonunique infinite estimates. At each iteration, the predicted probability for each observation to belong to its observed response group rapidly grows to 1 and the log likelihood diminishes to 0.

Quasicomplete Separation
: The maximum likelihood estimate does not exist. The data are not completely separated and there exists a vector of pseudoestimates that correctly allocates all but a nonempty subset of observations to their response groups. Such a data configuration also gives nonunique infinite estimates. The log likelihood does not diminish to 0 at each iteration, as in the case of complete separation.

Overlap
: If neither complete nor quasicomplete separation exists in the sample data, there is an overlap of sample points. The maximum likelihood estimate exists and is unique.

Complete separation and quasicomplete separation are typical problems for small samples. Complete separation can occur for any type of data, but quasicomplete separation is not likely for quantitative data. For more information on these three types of data configurations and more examples of how they can result in infinite parameters, see So (1993).

This example shows how the LOGISTIC procedure informs you (in Release 6.10 and later releases) when your model has infinite parameters due to complete or quasicomplete separation in your sample points. It also shows how to plot the data with the PLOT procedure and how to use the MEANS procedure to give you more information about the configuration of your sample points.

Program

Create the DIABETES data set. The complete DIABETES data set is in the Introduction.

```
data diabetes;
   input patient relwt glufast glutest instest sspg group;
   label relwt   = 'Relative weight'
         glufast = 'Fasting Plasma Glucose'
         glutest = 'Test Plasma Glucose'
         instest = 'Plasma Insulin during Test'
         sspg    = 'Steady State Plasma Glucose'
         group   = 'Clinical Group';
   datalines;
1  0.81  80  356 124   55  1
2  0.95  97  289 117   76  1
3  0.94 105  319 143  105  1
more data lines
;
```

Convert the ordinal response (GROUP) to a binary response (GRP). Arrange the input data so that the chemical diabetics and normals are combined into one group — the nonevent group. The overt diabetics are the event group.

```
data diabet3;
   set diabetes;
   grp=(group ne 3);
run;
```

Try to fit a logistic regression model to the revised diabetes data using GLUFAST as the explanatory variable.

```
proc logistic data=diabet3;
   model grp=glufast;
   title 'Diabetes Data';
run;
```

Plot the diabetes data to show the complete separation.

```
proc plot data=diabet3;
   plot glufast*patient=grp;
run;
```

Find the minimum and maximum values of GLUFAST for each group.

```
proc means data=diabet3 min max;
   by grp notsorted;
   var glufast;
run;
```

Create a data set with quasicomplete separation in the data points.

```
data diabet4;
   set diabet3;
   if glufast=114 then glufast=120;
run;
```

Try to fit a logistic regression model to the revised data.

```
proc logistic data=diabet4;
   model grp=glufast;
run;
```

Output

Output 11.1
PROC LOGISTIC Output Indicating Complete Separation

```
                              Diabetes Data

                          The LOGISTIC Procedure

     Data Set: WORK.DIABET3
     Response Variable: GRP
     Response Levels: 2
     Number of Observations: 145
     Link Function: Logit


                             Response Profile

                          Ordered
                            Value     GRP      Count

                                1       0         33
                                2       1        112


   WARNING: There is a complete separation in the sample points. The maximum
            likelihood estimate does not exist.
```

Output 11.2
Plot of Revised Diabetes Data Showing Complete Separation

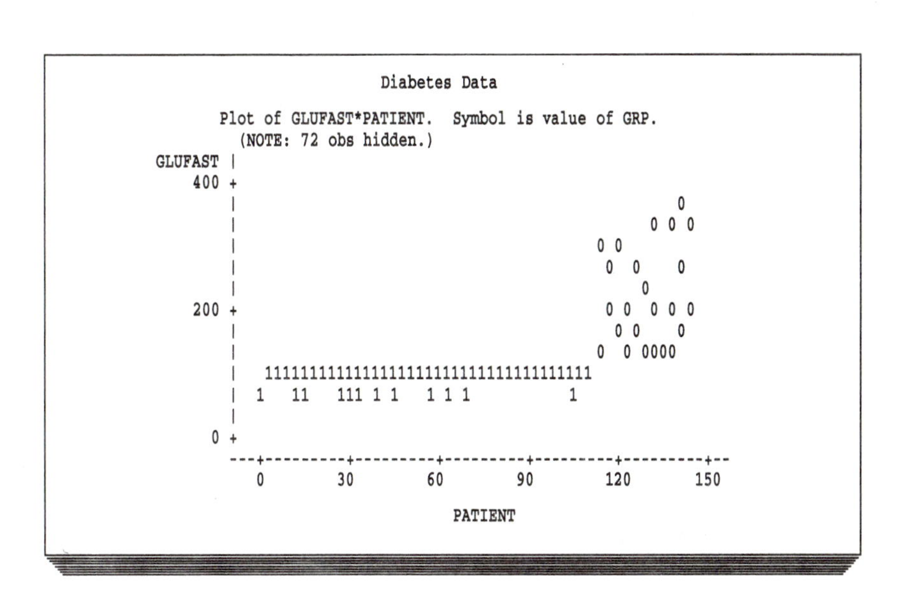

Output 11.3
Minimum and Maximum Values of the GLUFAST Variable for Each Group.

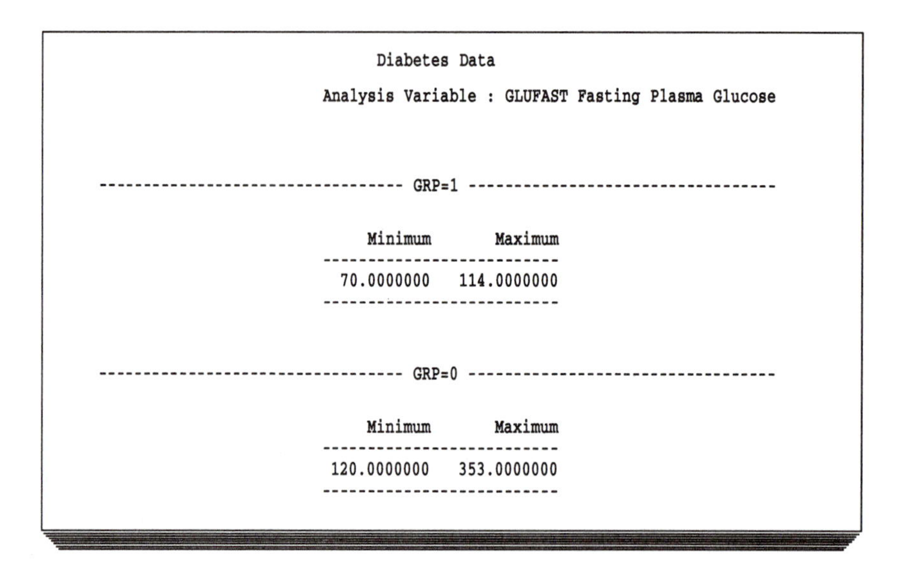

```
                          Diabetes Data

           Analysis Variable : GLUFAST Fasting Plasma Glucose


--------------------------------- GRP=1 ---------------------------------

                       Minimum       Maximum
                    --------------------------
                    70.0000000   114.0000000
                    --------------------------


--------------------------------- GRP=0 ---------------------------------

                       Minimum       Maximum
                    --------------------------
                    120.0000000   353.0000000
                    --------------------------
```

Output 11.4
PROC LOGISTIC Output Indicating Quasicomplete Separation

```
                          Diabetes Data

                     The LOGISTIC Procedure

     Data Set: WORK.DIABET4
     Response Variable: GRP
     Response Levels: 2
     Number of Observations: 145
     Link Function: Logit


                        Response Profile

                    Ordered
                      Value      GRP      Count

                          1        0         33
                          2        1        112

   WARNING: There is possibly a quasicomplete separation in the sample points.
            The maximum likelihood estimate may not exist.
```

Explanation

When complete separation exists in the input data, PROC LOGISTIC stops processing and prints a warning, as shown in Output 11.1. In this example, the independent variable, GLUFAST, perfectly predicts the response. Perfectly fitting models always have the problem of infinite parameters. In one sense, it is good to find a model that provides perfect prediction. However, PROC LOGISTIC cannot estimate the infinite parameters associated with a model that perfectly predicts the response.

Output 11.2 shows the GLUFAST variable plotted against the identifier variable, PATIENT, from the revised diabetes data, with each observation given the value of GRP (0 or 1). Note that all of the event observations have high values for GLUFAST, while the nonevent observations have lower values for GLUFAST.

Output 11.3 shows the minimum and maximum values of the GLUFAST variable for the event group and the nonevent group. The nonevent group, `GRP=1`, has values of GLUFAST ranging from 70 to 114. The event group, `GRP=0`, has values of GLUFAST ranging from 120 to 353. The results from

PROC PLOT and PROC MEANS confirm that there is complete separation between the two groups.

Output 11.4 shows the warning that PROC LOGISTIC prints when you have quasicomplete separation in the input data. In this case, a few observations in both of the two response groups share the value of 120 for the GLUFAST variable.

When you find that complete separation or quasicomplete separation exists in your sample points, you have various remedies:

- Examine the raw data for transcription errors.
- Categorize quantitative variables.
- Use fewer or different explanatory variables.
- Collect more data.
- Build a model up with forward or stepwise selection, if you encounter complete separation when you use the backwards elimination model selection method.
- Reclassify the response variable. In this example, the chemical diabetics are classified as nonevents. If they are reclassified as events, then the sample points have an overlap configuration, rather than a complete separation configuration.

When you use more explanatory variables in your model you increase the likelihood of encountering complete or quasicomplete separation.

Further Reading

- For complete reference information on the MEANS and PLOT procedures, see the *SAS Procedures Guide, Version 6, Third Edition.*

References

- Albert, A. and Anderson, J.A. (1984), "On the Existence of Maximum Likelihood Estimates in Logistic Regression Models," *Biometrika*, 71, 1-10.
- Hauck, W.W. and Donner, A. (1977), "Wald's Test As Applied to Hypotheses in Logit Analysis," *Journal of the American Statistical Association*, 72, 851-863.
- Santner, T.J. and Duffy, E.D. (1986), "A Note on A. Albert and J.A. Anderson's Conditions for the Existence of Maximum Likelihood Estimates in Logistic Regression Models," *Biometrika*, 73, 755-758.
- So, Y. (1993), "A Tutorial on Logistic Regression," *Proceedings of the Eighteenth Annual SAS Users Group International*, 1290-1295.

EXAMPLE 12

Logistic Regression with an Ordinal Response

Featured tools:

- PROC LOGISTIC
  - OUTEST= option
  - OUTPUT statement
- PROC GPLOT
- PROC CATMOD

Suppose you have a response variable that can take on any of *k* possible outcomes. If these outcomes can be ordered, then the response variable is measured on an ordinal scale. Typically, this ordering is a measure of degree, such as determination of disease status or strength of preference for an item. For example, the severity of coronary disease may be classified into three response categories: 1=no disease, 2=angina pectoris, and 3=myocardial infarction. The strength of response may take on values of strongly agree, agree, no opinion, disagree, and strongly disagree.

The LOGISTIC procedure can be used to fit a model with an ordinal response variable. It fits a parallel lines regression model that is based on the cumulative distribution probabilities of the response categories.

For example, suppose the response has three possible outcomes: 1=normal, 2=early disease state, and 3=advanced disease state. Define

$$p_1 = \text{Prob}(Y = 1 \mid X)$$
$$p_2 = \text{Prob}(Y = 2 \mid X)$$
$$p_3 = \text{Prob}(Y = 3 \mid X)$$

where Y is the response variable and X is a continuous predictor variable.

PROC LOGISTIC fits the model

$$\text{logit}(p_1) = \log\left(\frac{p_1}{1 - p_1}\right) = \alpha_1 + \beta x$$

$$\text{logit}(p_1 + p_2) = \log\left(\frac{p_1 + p_2}{1 - p_1 - p_2}\right) = \alpha_2 + \beta x$$

Notice that PROC LOGISTIC models the cumulative probabilities and that this model assumes a common slope parameter associated with the predictor variable. This model is known as the proportional-odds model because the ratio of the odds of the event $Y \leq j$ is independent of the category, *j*. In other words, the odds ratio is constant for all categories. If this assumption is not valid and you need to allow the odds ratio to change with respect to category, you need to fit a model with individual slope parameters. PROC CATMOD can fit such a model. See the section "Variations" for details.

Program

Set graphics options.

```
goptions cback=white colors=(black);
```

Create the DIABETES data set. The complete DIABETES data set is in the Introduction.

```
proc format;
   value gp 3='(3) Overt Diabetic ' 2='(2) Chem. Diabetic' 1='(1) Normal';
run;

data diabetes;
   infile datalines eof=endfile;
   input patient relwt glufast glutest instest sspg group;
   label relwt   = 'Relative weight'
         glufast = 'Fasting Plasma Glucose'
         glutest = 'Test Plasma Glucose'
         instest = 'Plasma Insulin during Test'
         sspg    = 'Steady State Plasma Glucose'
         group   = 'Clinical Group';
   endfile:  do glutest=250 to 1600 by 25;
                output;
             end;
   datalines;
  1  0.81  80  356 124   55  1
  2  0.95  97  289 117   76  1
  3  0.94 105  319 143  105  1

  <more data lines>

143  0.90 213 1025  29  209  3
144  1.11 328 1246 124  442  3
145  0.74 346 1568  15  253  3
;
```

Fit the logistic regression model. Use the descending option to model the probability of being diabetic.

```
proc logistic data=diabetes descending;
   model group=glutest;
   output out=probs predicted=prob xbeta=logit;
   format group gp.;
run;
```

Print the first ten observations of PROBS.

```
proc print data=probs (obs=10);
   var patient group glutest _level_ prob;
   title1 'Printout of first 10 observations of PROBS';
run;
```

Reset the TITLE1 statement.

```
title1;
```

Sort GLUTEST before plotting.

```
proc sort data=probs;
   by glutest;
run;
```

Plot the parallel logit and probability curves.

```
proc gplot data=probs;
   plot logit*glutest=_level_ / frame;
   title2 'Plot of Parallel Logits';
run;
   plot prob*glutest=_level_ / frame;
   title2 'Plot of Parallel Probability Lines';
run;
```

Create data sets containing predicted values for the overt and chemical groups. Use these to plot the individual predicted curves.

```
data overt(rename=(prob=p3)) chem(rename=(prob=p2));
   set probs;
   if mod(_n_,2)=1 then output overt;
   else output chem;
run;
```

Compute the individual probabilities of group membership.

```
data probs2;
   merge overt chem;
   keep glutest prob_3 prob_2 prob_1;
   prob_3=p3;
   prob_2=p2-p3;
   prob_1=1-p2;
run;
```

Sort PROBS2 before plotting.

```
proc sort data=probs2;
   by glutest;
run;
```

Define a plotting symbol for each outcome.

```
symbol1 i=join line=1 value=x height=.75;
symbol2 i=join line=2 value=plus height=.75;
symbol3 i=join line=3 value=circle height=.75;
```

Label the axes.

```
axis1 label=(angle=-90 rotate=90 'Probability');
axis2 label=('Test Plasma Glucose');
```

Title the plot.

```
title2 'Individual Density Plots';
```

Add a footnote to serve as a legend for the plotting symbols.

```
footnote1
   height=.75 font=zapf 'x x x ' h=1.0 font=zapf ' Overt '
   height=.75 font=zapf '+ + + ' h=1.0 font=zapf ' Chemical '
   height=.75 font=special 'H H H ' h=1.0 font=zapf ' Normal';
```

Plot the probability functions.

```
proc gplot data=probs2;
   plot prob_3*glutest=1
        prob_2*glutest=2
        prob_1*glutest=3
        / overlay vaxis=axis1 haxis=axis2 frame;
run;
```

Output

Output: PROC LOGISTIC

Output 12.1
PROC LOGISTIC Output for Ordinal Response Regression Model

```
                          The LOGISTIC Procedure

   Data Set: WORK.DIABETES
   Response Variable: GROUP     Clinical Group
   Response Levels: 3
   Number of Observations: 145
   Link Function: Logit


                             Response Profile

                   Ordered
                     Value  GROUP                   Count

                         1  (3) Overt Diabetic         33
               ❶         2  (2) Chem. Diabetic         36
                         3  (1) Normal                 76

WARNING: 55 observation(s) were deleted due to missing values for the
         response or explanatory variables.


           ❷     Score Test for the Proportional Odds Assumption

               Chi-Square = 16.0881 with 1 DF (p=0.0001)


                    Testing Global Null Hypothesis: BETA=0

                                 Intercept
                     Intercept      and
   Criterion           Only      Covariates    Chi-Square for Covariates

   AIC               300.198       48.358           .
   SC                306.152       57.289           .
❸  -2 LOG L          296.198       42.358        253.840 with 1 DF (p=0.0001)
   Score                .             .           92.434 with 1 DF (p=0.0001)


                    Analysis of Maximum Likelihood Estimates
                ❹        ❺          ❻          ❼
              Parameter Standard     Wald       Pr >    Standardized     Odds
  Variable DF  Estimate   Error   Chi-Square Chi-Square   Estimate      Ratio

  INTERCP1 1   -41.0995   8.3690    24.1172    0.0001           .          .
  INTERCP2 1   -29.1971   5.9300    24.2425    0.0001           .          .
  GLUTEST  1     0.0691   0.0142    23.5179    0.0001   12.070608      1.072


          Association of Predicted Probabilities and Observed Responses

                   Concordant = 64.0%         Somers' D = 0.639
                   Discordant =  0.1%         Gamma     = 0.997
                   Tied       = 35.9%         Tau-a     = 0.394
                   (6432 pairs)               c         = 0.819
```

Explanation

❶ `Ordered Value` gives the sort order of the response levels. Be sure to verify that the response levels are ordered correctly. With an ordinal response, the response levels should be sorted in either ascending or descending order. In this analysis, the model predicts the probability of more disease.

❷ `Score Test` gives the chi-square test for the proportional odds assumption. When this assumption is not reasonable, you should consider fitting a model with distinct slope parameters rather than a common one.

The small p-value for this test indicates that the proportional odds assumption may not be reasonable for these data. You may still want to consider investigating a model that allows separate slopes for the two cumulative logit response functions.

However, Peterson and Harrell (1990) showed that this test is very anti-conservative. They recommend using this test only to conclude that the proportional odds assumption is valid (based on a large p-value).

❸ `-2 LOG L` gives the -2 log likelihood for the model. Since there is a single predictor variable in the model, this also serves as a likelihood ratio chi-square test for GLUTEST. Here you see that GLUTEST is important ($p = 0.0001$) in predicting group membership.

❹ `Parameter Estimate` gives the parameter estimates. Defining $\hat{p}_3$ as the probability of being overt diabetic, $\hat{p}_2$ as the probability of being chemical diabetic, and $\hat{p}_1$ as the probability of being normal, the two (parallel) regression lines fitted are

$$\text{logit}\left(\hat{p}_3\right) = -41.0995 + 0.0691 \times \text{GLUTEST}$$

$$\text{logit}\left(\hat{p}_2 + \hat{p}_3\right) = -29.1971 + 0.0691 \times \text{GLUTEST}$$

The positive coefficient for GLUTEST indicates that as GLUTEST increases, both response functions increase. This means that higher levels of GLUTEST are associated with increased probability of being diabetic.

❺ `Standard Error` gives the standard errors for the parameter estimates.

❻ `Wald Chi-Square` is the Wald chi-square statistic for each parameter in the model. This statistic is the square of the ratio of the parameter estimate to its standard error.

❼ `Pr>Chi-Square` is the significance level of the Wald chi-square test statistic. Notice that all parameters are highly significant when compared to a typical cut-off value of 0.05.

Output: PROC PRINT

Output 12.2
Printout of the First 10 Observations of the PROBS Data Set

Printout of first 10 observations of PROBS

OBS	PATIENT	GROUP	GLUTEST	_LEVEL_	PROB
1	1	(1) Normal	356	(3) Overt Diabetic	.0000001
2	1	(1) Normal	356	(2) Chem. Diabetic	.0098922
3	2	(1) Normal	289	(3) Overt Diabetic	.0000000
4	2	(1) Normal	289	(2) Chem. Diabetic	.0000976
5	3	(1) Normal	319	(3) Overt Diabetic	.0000000
6	3	(1) Normal	319	(2) Chem. Diabetic	.0007750
7	4	(1) Normal	356	(3) Overt Diabetic	.0000001
8	4	(1) Normal	356	(2) Chem. Diabetic	.0098922
9	5	(1) Normal	323	(3) Overt Diabetic	.0000000
10	5	(1) Normal	323	(2) Chem. Diabetic	.0010214

Explanation

Output 12.2 is a listing of the first ten observations of PROBS. Notice that there are two observations for each patient. The observation with **`_LEVEL_=(3) Overt Diabetic`** gives the estimated probability that a patient is overt diabetic. The observation with **`_LEVEL_=(2) Chem. Diabetic`** gives the estimated probability that a patient is either overt or chemical diabetic.

For example, the estimated probability that patient 1 (who has a GLUTEST equal to 356 and is normal) is overt diabetic is 0.0000001, while the probability that patient 1 is chemical or overt diabetic is 0.0098922.

Output: Plot of Parallel Logits

Output 12.3
Output from PROC GPLOT

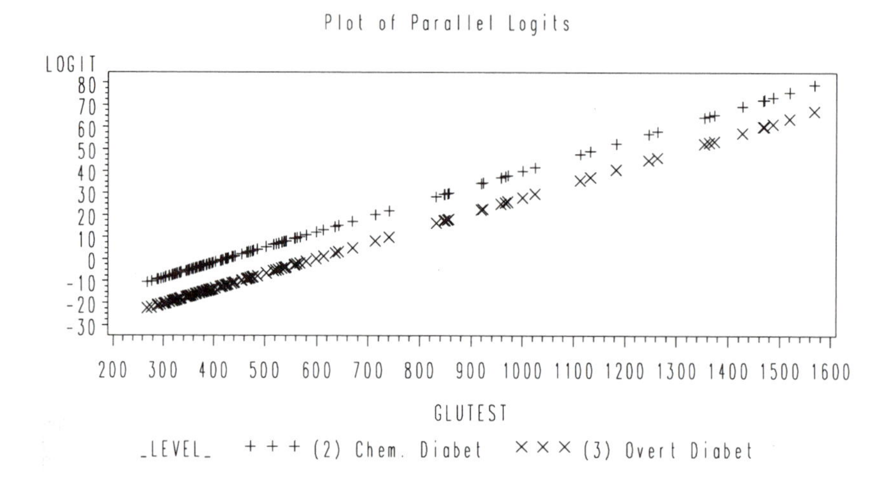

Explanation

Output 12.3 displays the plots of the two cumulative logit response functions. The proportional odds assumption forces the lines to be parallel, as they have a common slope parameter.

Output: Plot of Probability Curves

Output 12.4
Output from PROC GPLOT

Plot of Parallel Probability Lines

PROB

1.0 0.9 0.8 0.7 0.6 0.5 0.4 0.3 0.2 0.1 0.0

200 300 400 500 600 700 800 900 1000 1100 1200 1300 1400 1500 1600

GLUTEST

LEVEL + + + (2) Chem. Diabet × × × (3) Overt Diabet

Explanation

Output 12.4 displays the plots of the predicted probabilities curves as derived from the cumulative logit response functions. The rightmost curve is given by

$$\hat{p}_3 = \frac{e^{-41.0995 + 0.0691 \times \text{GLUTEST}}}{1 + e^{-41.0995 + 0.0691 \times \text{GLUTEST}}}$$

The leftmost curve is given by

$$\left(\hat{p}_3 + \hat{p}_2\right) = \frac{e^{-29.1971 + 0.0691 \times \text{GLUTEST}}}{1 + e^{-29.1971 + 0.0691 \times \text{GLUTEST}}}$$

Output: Plot of Individual Densities

Output 12.5
Output from PROC GPLOT

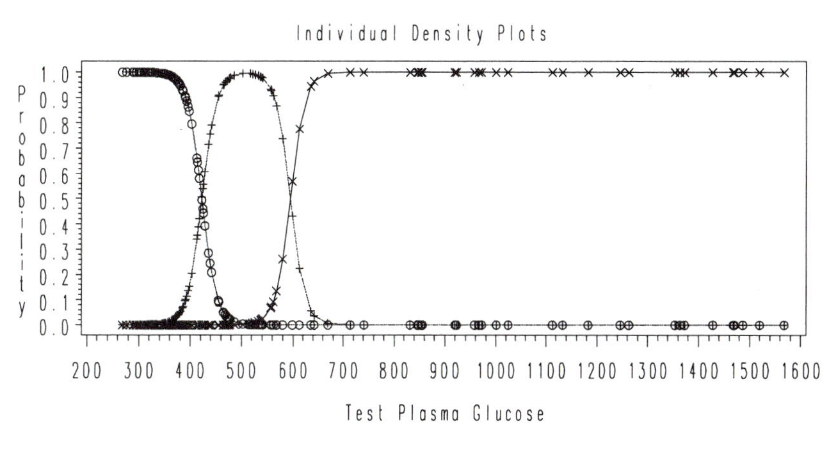

Explanation

Output 12.5 shows the individual probability densities for each group. These are given by

$$\hat{p}_3 = \frac{e^{-41.0995 + 0.0691 \times \text{GLUTEST}}}{1 + e^{-41.0995 + 0.0691 \times \text{GLUTEST}}}$$

$$\hat{p}_2 = \frac{e^{-29.1971 + 0.0691 \times \text{GLUTEST}}}{1 + e^{-29.1971 + 0.0691 \times \text{GLUTEST}}} - \hat{p}_3$$

$$\hat{p}_1 = 1 - \hat{p}_3 - \hat{p}_2$$

You can determine the points where the curves in Output 12.5 intersect by setting $\hat{p}_1 = \hat{p}_2$ and $\hat{p}_2 = \hat{p}_3$. These points give you the estimated cutpoints for assigning patients to a group. The intersection points are given by

$$\text{C1} = \log\left(\frac{e^{\alpha_2} - 2e^{\alpha_1}}{e^{\alpha_1 + \alpha_2}}\right) / \beta$$

$$\text{C2} = \log\left(\frac{1}{e^{\alpha_2} - 2e^{\alpha_1}}\right) / \beta$$

In this case, the cutpoints are the GLUTEST values of 422 and 595. This means that if GLUTEST is less than 422, the model predicts normal; if GLUTEST is greater than 595, the model predicts overt diabetic; if in between, the model predicts chemical diabetic.

A Closer Look

Create the DIABETES Data Set

The data set to create the DIABETES data set not only reads the observed values but also outputs a set of observations with missing values for GROUP. These observations are used later for plotting the predicted probability curves. They are not used in estimating the model parameters.

`infile datalines eof=endfile`
: does two things. First, `datalines` specifies that the input data immediately follow the DATALINES statement. Second, `eof=endfile` jumps to the labeled statement when end-of-file is reached.

DO loop
: generates and outputs values of GLUTEST ranging from 250 to 1600 in increments of 25. These observations have a missing value for the response variable GROUP. Although PROC LOGISTIC does not use these observations to fit the model, it does calculate predicted values for them. You use these data points to fill in the plot of the fitted probit curve using PROC GPLOT.

Create Data Sets Containing Predicted Values for the Overt and Chemical Groups

The OUTPUT statement creates an output data set that contains the predicted probabilities of group membership. The output data set contains a pair of observations for each observation in the input data set. The first observation in the pair contains the predicted probability of being overt diabetic. The second contains the predicted probability of being either overt or chemical diabetic. You can use the MOD function to create separate data sets for the pairs. You then need to rename variables and merge them back together.

Variations

Computing the Predicted Probabilities in a DATA Step

Suppose you want to use the logistic regression equation to compute predicted probabilities for a new set of observations. For example, you may want to generate dummy observations with values of GLUTEST that were not observed in the actual data and then plot a smooth predicted probability curve. In that case, you can create an output data set that contains the parameter estimates and then use a DATA step to compute the predicted probabilities.

Fit the logistic model. Use the OUTEST= option to create an output data set that contains the parameter estimates.

```
proc logistic data=diabetes descending outest=parms;
   model group=glutest;
   format group gp.;
run;
```

Rename GLUTEST.

```
data parms;
   set parms;
   rename glutest=b1;
run;
```

Use the logistic regression equations to compute the probability for each response.

Use a DO loop to generate values over the range of values of GLUTEST. Compute the values of the two logit functions. Then exponentiate the logits to evaluate the cumulative probabilities.

```
data parms2;
   set parms;

   do glutest=250 to 1600 by 25;

      logit1=intercp1 + b1*glutest;
      logit2=intercp2 + b1*glutest;

      p1=exp(logit1)/(1+exp(logit1));
      p2=exp(logit2)/(1+exp(logit2));
```

Compute the probability for each response outcome and output the observation.

```
      prob_1=p1;
      prob_2=p2-p1;
      prob_3=1-p2;
      output;
   end;
run;
```

Overlay the plots of the predicted probabilities.

```
proc gplot data=parms2;
   plot prob_3*glutest
        prob_2*glutest
        prob_1*glutest
        / overlay frame;
run;
```

Fitting a Model with Separate Slope Parameters

Fitting a Cumulative Logit Model with PROC CATMOD

If the common slope assumption of the cumulative logistic model with proportional odds is not reasonable, you can use PROC CATMOD to fit a model with separate slope parameters for the two cumulative logit response functions. The following code does this:

```
proc catmod data=diabetes;
   direct glutest;
   response clogits;
   model group=glutest;
```

CAUTION!

PROC CATMOD fits this model using weighted least squares rather than maximum likelihood. You may have problems fitting the model due to sparseness of data when you have a continuous response, as you do in this example. If PROC CATMOD cannot fit the model, it prints a message to the log. ■

For the DIABETES data set, you get the following message printed to the log:

```
      proc catmod data=diabetes;
         direct glutest;
         response clogits ;
         model group = glutest;
      run;

   ERROR: The response functions are linearly dependent since the number of
          functions per population, 2, is greater than or equal to the number
          of response levels, 1, in population 1.
```

The message indicates that you have populations with zero cell counts. One way to deal with this problem is to categorize the continuous covariate (in this case, GLUTEST) to eliminate zero cell counts and run the code again.

Fitting a Generalized Logit Model with PROC CATMOD

One more possible model is the generalized logit model given by

$$\log\left(\frac{p_1}{p_3}\right) = \alpha_1 + \beta_1 x$$

$$\log\left(\frac{p_2}{p_3}\right) = \alpha_2 + \beta_2 x$$

The following code fits this model and creates an output data set named CAT_PROB that contains the predicted probabilities of group membership.

```
proc catmod data=diabetes;
   direct glutest;
   response logits / out=cat_prob;
   model group=glutest;
run;
```

Output 12.6
Partial Output from PROC CATMOD

```
              MAXIMUM-LIKELIHOOD ANALYSIS-OF-VARIANCE TABLE

            Source                  DF   Chi-Square      Prob
            --------------------------------------------------
            INTERCEPT                2        13.57    0.0011
            GLUTEST                  2        12.70    0.0017

            LIKELIHOOD RATIO       232        26.47    1.0000


                  ANALYSIS OF MAXIMUM-LIKELIHOOD ESTIMATES

                                              Standard    Chi-
     Effect             Parameter  Estimate     Error    Square   Prob
     ----------------------------------------------------------------
     INTERCEPT                  1     109.8   43.0212      6.52 0.0107
                                2   19.4528    6.4615      9.06 0.0026
     GLUTEST                    3   -0.2471    0.1018      5.90 0.0152
                                4   -0.0319    0.0111      8.18 0.0042
```

Explanation

The fitted model is given by

$$\log\left(\frac{\hat{p}_1}{\hat{p}_3}\right) = 109.8 - .2471 \times \text{GLUTEST}$$

$$\log\left(\frac{\hat{p}_2}{\hat{p}_3}\right) = 19.4528 - .0319 \times \text{GLUTEST}$$

The parameter estimates are not comparable in any sense to those fitted by PROC LOGISTIC because the models are different. However, you can compare the predicted probabilities of these two models. The following table presents the predicted probabilities for two patients.

	Observed Proportions			Predicted Probabilities					
				PROC LOGISTIC			PROC CATMOD		
GLUTEST	Normal	Chemical	Overt	Normal	Chemical	Overt	Normal	Chemical	Overt
360	1.0	0.0	0.0	0.987	0.013	0.0	1.0	0.0	0.0
426	0.5	0.5	0.0	0.4429	0.5571	0.00001	0.2098	0.7880	0.0022
599	0.0	1.0	0.0	0.0	0.4312	0.5688	0.0	0.5910	0.4090
636	0.0	0.0	1.0	0.0	0.0556	0.9444	0.0	0.3078	0.6922
643	0.0	1.0	0.0	0.0	0.0350	0.9650	0.0	0.2624	0.7376
714	0.0	0.0	1.0	0.0	0.0003	0.9997	0.0	0.0357	0.9643

Reference

Peterson, B. and Harrell, F. (1990), "Partial proportional odds models for ordinal response variables," *Applied Statistics*, 39, 205-217.

EXAMPLE 13

Analysis of 1:1 Matched Data

Featured tools:

- DATA step programming
- PROC LOGISTIC
  - MODEL statement, NOINT option
  - OUTPUT statement
  - TEST statement
- PROC GPLOT
  - PLOT statement
  - BUBBLE statement

A matched case-control study has two groups: a *case* group, composed of subjects who exhibit the outcome under investigation, and a *control* group, composed of subjects who do not exhibit the outcome. Matching is a way to balance the two groups with respect to one or more risk factors that are either known or thought to be associated with the outcome; that is, matching constrains the control group to be similar to the case group with respect to the matching variables.

In a 1:1 matched study, each subject in the case group is matched, or paired, with exactly one subject in the control group. The matching is accomplished by selecting a subject from the control group who is similar to a subject in the case group on the basis of one or more matching variables. For example, suppose age and sex are the matching variables. If a case is 25 years old and female, then a 25 year-old female control needs to be paired with this case.

Analysis of this kind of data requires conditional logistic regression, in which you condition on the total number of cases and the number of subjects in each matched set. In the case of 1:1 matching, the likelihood function for the conditional logistic model can be fit using PROC LOGISTIC by setting the response equal to a constant. You also need to manipulate your data so that the data vector contains the differences between the covariates of the case and the control for each pair.

In this example, you analyze from a large study investigating risk factors related to low birth weight. This data set consists of 56 age-matched pairs.

Program

Set graphics options.

```
goptions cback=white colors=(black);
```

Create the MATCH_11 data set. The complete data set is in the Introduction.

```
data match_11;
   input pair low age lwt race smoke ptd ht ui;
   datalines;
 1  0 14 135 1 0 0 0 0
 1  1 14 101 3 1 1 0 0
 2  0 15  98 2 0 0 0 0
 2  1 15 115 3 0 0 0 1
 3  0 16  95 3 0 0 0 0
 3  1 16 130 3 0 0 0 0

 more data lines

56  0 34 170 1 0 1 0 0
56  1 34 187 2 1 0 1 0
;
```

Create the data set of differences. Each observation contains the differences between covariates of each pair.

```
data diffs;
   set match_11;
```

Initialize temporary variables to hold covariate values for the control of each pair.

```
   retain pair1 lwt1 race3 race4 smoke1 ptd1 ht1 ui1  0;
   drop pair1 lwt1 race race3 race4 smoke1 ptd1 ht1 ui1;
```

Create indicator variables for race.

```
   select(race);
      when (1) do;
         race1=0;
         race2=0;
      end;
      when (2) do;
         race1=1;
         race2=0;
      end;
      when (3) do;
         race1=0;
         race2=1;
      end;
   end;
```

Copy the values of the covariates of each control into temporary variables.

```
   if (pair ne pair1) then do;
      pair1=pair;
      lwt1=lwt;
      race3=race1;
      race4=race2;
      smoke1=smoke;
      ptd1=ptd;
      ht1=ht;
      ui1=ui;
   end;
```

Calculate the differences (case-control) between the covariates of each pair and output the observation.

```
   else do;
      lwt=lwt-lwt1;
      race1=race1-race3;
      race2=race2-race4;
      smoke=smoke-smoke1;
      ptd=ptd-ptd1;
      ht=ht-ht1;
      ui=ui-ui1;
      output;
   end;
```

Fit the model with PROC LOGISTIC. Specify LOW as the outcome variable. Use NOINT to fit a model without an intercept term. Use TEST to test the overall effect of race.

```
title1 'Analysis of 1:1 Matched Case-Control Data';
proc logistic data=diffs;
   model low=race1 race2 smoke ht ui ptd lwt / noint;
   race: test race1=0, race2=0;
```

Create an output data set of diagnostic statistics. Request that influence statistics and the predicted probability be output.

```
   output out=stats
          difchisq=d_chi
          difdev=d_dev
          h=hat
          predicted=pred;
run;
```

Reset TITLE1 and disable the labels.

```
title1;
options nolabel;
```

Plot the diagnostic statistics against the predicted probability.

```
proc gplot data=stats;
   axis1 label=(angle=-90 rotate=90 'Deviance Change');
   plot d_dev*pred / frame vaxis=axis1;
   title2 'Change in Deviance Versus Predicted Probability';
run;
```

Produce a bubble plot using the influence (HAT) as the bubble size.

```
   axis2 label=(angle=-90 rotate=90 'Chi square Change');
   bubble d_chi*pred=hat / frame vaxis=axis2;
   title2 'Change in Chi-Square Versus Predicted Probability';
run;
```

Output

Ouput: PROC LOGISTIC

Output 13.1
PROC LOGISTIC Output for the 1:1 Matched Case-Control Data

```
                 Analysis of 1:1 Matched Case-Control Data

                        The LOGISTIC Procedure

  Data Set: WORK.DIFFS
  Response Variable: LOW
  Response Levels: 1
  Number of Observations: 56
  Link Function: Logit


                          Response Profile

                     Ordered
                       Value      LOW      Count

                           1        1         56


              Testing Global Null Hypothesis: BETA=0

                 Without        With
Criterion      Covariates    Covariates    Chi-Square for Covariates

AIC               77.632        66.946           .
SC                77.632        81.123           .
❶ -2 LOG L         77.632        52.946        24.687 with 7 DF (p=0.0009)
Score                .             .           19.492 with 7 DF (p=0.0068)


              Analysis of Maximum Likelihood Estimates
                  ❷                     ❸          ❹                       ❺
               Parameter Standard     Wald        Pr >    Standardized     Odds
Variable  DF   Estimate    Error   Chi-Square Chi-Square    Estimate       Ratio

RACE1      1     0.6174   0.6851     0.8121     0.3675      0.199958      1.854
RACE2      1    -0.0214   0.6888     0.0010     0.9752     -0.009138      0.979
SMOKE      1     1.3485   0.6116     4.8618     0.0275      0.516078      3.852
HT         1     2.2104   1.0437     4.4852     0.0342      0.512156      9.119
UI         1     1.2370   0.7243     2.9165     0.0877      0.345611      3.445
PTD        1     1.7826   0.7754     5.2854     0.0215      0.543821      5.946
LWT        1    -0.0169  0.00975     3.0066     0.0829     -0.433691      0.983

NOTE: Measures of association between the observed and predicted values
      were not calculated because the predicted probabilities are
      indistinguishable when they are classified into intervals of length
      0.002 .

                ❻   Linear Hypotheses Testing

                          Wald                               Pr >
          Label        Chi-Square            DF         Chi-Square
          RACE             1.0257             2             0.5988
```

Explanation

❶ **`-2 LOG L`** gives the likelihoods for the model without covariates, for the model with the covariates, and the ratio chi-square statistic for the test that at least one covariate has a nonzero parameter. This statistic is large enough (that is, the *p*-value is small enough) to conclude that the full model is significantly different from the null model.

❷ **`Parameter Estimate`** is the estimated parameter (relative log odds) associated with each covariate. A positive parameter estimate indicates that the likelihood of a low birth weight baby increases with the presence (or level) of the covariate, while a negative parameter estimate indicates that the likelihood decreases with the presence of the covariate.

CAUTION!

Changing the order of the way differences between the cases and controls are taken would reverse the signs and the interpretation. ■

❸ **`Wald Chi-Square`** is the Wald statistic for the significance of the covariate in the fitted model. Use the associated *p*-value to determine the importance of each covariate.

❹ **`Pr>Chi-Square`** is the *p*-value for the Wald statistic. Small values indicate that the covariate is important in predicting the outcome event. Notice that all covariates except RACE1 and RACE2 are statistically significant at the 0.10 level.

❺ **`Odds Ratio`** is the odds ratio for the covariate. The odds ratios are the exponentiated parameter estimates. They provide an estimate of the expected change in the odds of a mother having a low birth weight baby versus a mother having a normal weight baby for a unit increase in the covariate.

In this example, the variable SMOKE is a binary predictor variable, coded so that SMOKE=1 for mothers who smoked during pregnancy and SMOKE=0 for mothers who didn't. Therefore, the difference between a smoking and a nonsmoking mother represents a one unit change. (Be careful, as other coding schemes may code a binary predictor variable with the values 1 and −1.) The odds ratio for SMOKE is 3.852. This means that the odds for smoking mothers having a low birth weight baby are 3.852 times larger than the odds for nonsmoking mothers. Smoking mothers are nearly four times more likely to have low birth weight babies.

❻ **`Linear Hypothesis Testing`** gives the results of the TEST statement. The overall race effect is tested by testing that the two parameters for the two dummy variables, RACE1 and RACE2, are jointly equal to zero. The large *p*-value for this test indicates that race has little association with the response variable. This test uses the Wald chi-square statistic.

Output: Scatter Plot

Output 13.2
Scatter Plot of the Change in Deviances Vs. the Predicted Probabilities

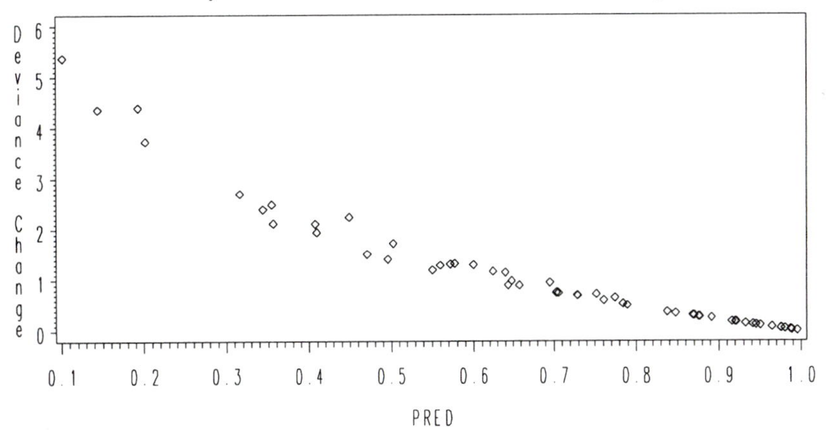

Explanation

Output 13.2 shows the change in the deviance, D_DEV, due to deleting each matched pair, plotted against the predicted probability of an event. Notice that D_DEV increases as the predicted probability decreases. Also notice that four points have relatively large values. This indicates that these four pairs are poorly fit by the model.

Output: Bubble Plot

Output 13.3
Bubble Plot Where the Bubble Size Is Proportional to the HAT Statistic

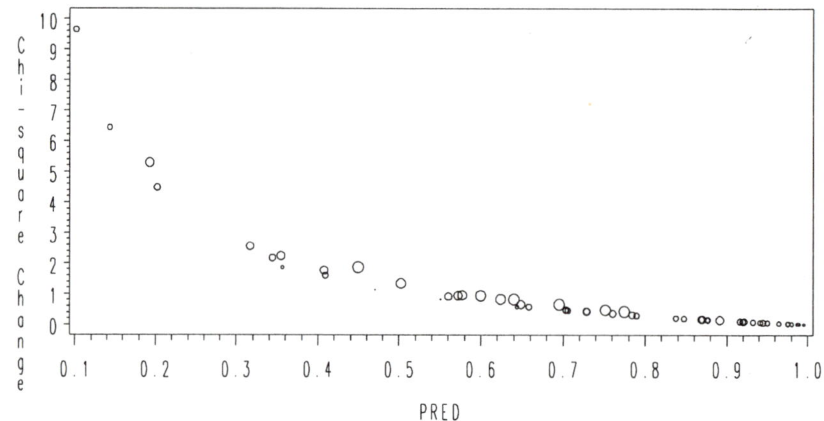

Explanation

Output 13.3 shows a bubble plot of the change in the chi-square goodness-of-fit test, D_CHI, plotted against the predicted probabilities. The size of the bubble is proportional to HAT, the leverage of the point in the design space. Notice here that only one of the four points that are poorly fit has a relatively large leverage.

A Closer Look

Create the Data Set of Differences

The DATA step code to create DIFFS outputs one observation for each pair, with the values for the covariates being the differences between the case and the control of each pair. This is accomplished efficiently in a single pass through the data.

For this example, the control of each pair immediately precedes the case. When the DATA step is executed

1. Temporary variables, PAIR1—UI1, are initialized to zero with a RETAIN statement. These variables are eventually dropped from the data set.
2. A SELECT statement and DO groups create the two indicator variables for race.
3. When the data for a control are processed, PAIR and PAIR1 are unequal. This causes the values for the covariates to be copied to the temporary

variables. In addition, PAIR1 is set equal to PAIR. Note that no observations are output in this DO group.

4. When the data for a case are processed, PAIR is equal to PAIR1. The temporary variables now hold the values for the controls covariates so that taking the difference is possible. The values for the covariates of the case are replaced with the differences. This observation is then output to data set DIFFS.

Create an Output Data Set of Diagnostic Statistics

PROC LOGISTIC can calculate and output several diagnostic statistics that you can use to assess the adequacy of the fitted model. You can use DIFCHISQ and DIFDEV to identify observations that contribute heavily to the disagreement between the data and the predicted values of the fitted model. These statistics are based on the change in the fitted model when an observation is excluded. Because each observation in the data set represents the pair (case and control), omitting a data point is equivalent to omitting the entire pair. The following influence statistics are requested in this example:

- DIFCHISQ is the change in the chi-squared goodness-of-fit statistic attributable to deleting the individual observation.
- DIFDEV is the change in the deviance attributable to deleting the individual observation.
- H is the leverage (the diagonal element of the hat matrix) for detecting extreme points in the design space.
- PREDICTED is the predicted probability of an event.

References

SAS Institute Inc. (1990), *SAS/STAT User's Guide, Version 6, First Edition, Volume 2*, Cary, NC: SAS Institute Inc.

Hosmer, D. W. and Lemeshow, S. (1989), *Applied Logistic Regression*, NY: John Wiley & Sons Inc.

Kleinbaum, D. G. (1994), *Logistic Regression: A Self-Learning Text*, NY: Springer-Verlag.

EXAMPLE 14

Analysis of N:M Matched Data

Featured tools:

- DATA step programming
- PROC PHREG

  MODEL statement, TIES= option

  STRATA statement

You can use the PHREG procedure to fit a conditional logistic regression model to investigate the relationship between an outcome and a set of risk factors in matched case-control studies. In Example 13, a 1:1 matched case-control study was analyzed. You can also have multiple controls matched to each case, a 1:M matching design, or have a varying number of cases and controls matched, an N:M matching design.

While PROC PHREG is intended to model survival times, with censored observation allowed, you can modify your matched data so that PROC PHREG fits a conditional logistic model. To do this, use PROC PHREG and form a stratum for each matched set. On the MODEL statement, specify TIES=BRESLOW if you have 1:M matched data, and TIES=DISCRETE if you have N:M matched data. You also need to create a survival time variable so that all cases in a matched set, or stratum, have the same event time value and so that corresponding controls are censored at later times.

In this example, you analyze data from a hospital-based case-control study designed to investigate risk factors associated with benign breast disease. Data are provided on 50 cases and 150 age-matched controls, with three controls per case. However, there are observations with missing values for some risk factors. This tends to complicate the analysis. Two strategies for analyzing this data are

- Use only matched sets where the case and three controls have complete data. For studies where the matched sets with missing data are few, this approach is reasonable. However, in this example, only 29 of 50 matched sets are available for the analysis, meaning that you would discard over 40 percent of the data. See the section, "Variation," for the code to take this approach when there are relatively few missing data.
- Use all matched sets but analyze using N:M matching based on age. This discards the 1:M matching design for the general design but enables you to examine all risk factors. However, because there are some controls who no longer are matched with cases, these observations do not contribute to the analysis.

Program

Create the MATCH_NM data set. The complete MATCH_NM data set is in the Introduction.

```
data match_NM;
   input matchset subject agmt fndx chk agp1 agmn nlv liv  wt  aglp;
   datalines;
 1  1   39  1   1 23   13  0  5 118  39
 1  2   39  0   0 16   11  1  3 175  39
 1  3   39  0   0 20   12  1  3 135  39
 1  4   39  0   1 21   11  0  3 125  40
 2  1   38  1   1  .   14  .  . 118  39
 2  2   38  0   1 20   15  0  2 183  38
 2  3   38  0   1 19   11  0  5 218  38
```

```
 2  4  38  0   1 23   13  0  2 192  37
 3  1  38  1   1 22   15  2  2 125  38
 3  2  38  0   1 20   14  0  2 123  38
 3  3  38  0   1 19   13  3  2 140  37
 3  4  38  0   1 18   13  0  2 160  38

 more data lines

50  1  41  1  1 34   13  1  2 138  42
50  2  42  0  1  .   13  .  . 118  41
50  3  41  0  0 30   12  1  2 129  41
50  4  41  0  1 21   12  0  2 180  41
;
```

Create an event time variable. STATUS is set equal to 1 for cases and equal to 2 for controls.

```
data match_NM;
   set match_NM;
   status=2-fndx;
run;
```

Fit the model.
Use the TIES=DISCRETE option and stratify on age, AGMT.

```
proc phreg data=match_NM;
   model status*fndx(0)=chk agp1 agmn nlv liv wt aglp / ties=discrete;
   strata agmt;
run;
```

Output

Output 14.1
Output from PROC PHREG Using All Matched Sets

The PHREG Procedure

Data Set: WORK.MATCH_NM
Dependent Variable: STATUS
Censoring Variable: FNDX
Censoring Value(s): 0
Ties Handling: DISCRETE

❶ Summary of the Number of Event and Censored Values

Stratum	AGMT	Total	Event	Censored	Percent Censored
1	27	1	0	1	100.00
2	28	2	0	2	100.00
3	30	1	0	1	100.00
4	31	3	1	2	66.67
5	32	5	1	4	80.00
6	33	6	1	5	83.33
7	34	3	1	2	66.67
8	35	8	2	6	75.00
9	36	7	0	7	100.00
10	37	3	0	3	100.00
11	38	18	4	14	77.78
12	39	4	1	3	75.00
13	40	1	0	1	100.00
14	41	10	3	7	70.00
15	42	0	0	0	.
16	43	4	1	3	75.00
17	44	7	2	5	71.43
18	45	15	3	12	80.00
19	46	8	2	6	75.00
20	47	8	2	6	75.00
21	48	3	1	2	66.67
22	49	3	0	3	100.00
23	51	4	1	3	75.00
24	52	7	1	6	85.71
25	53	3	1	2	66.67
26	55	6	2	4	66.67
27	56	4	1	3	75.00
28	58	3	1	2	66.67

```
      29    60              4          1          3       75.00
      30    61             12          4          8       66.67
      31    62              7          0          7      100.00
      32    63              2          1          1       50.00
      33    64              3          1          2       66.67
      34    68              3          1          2       66.67
    ---------------------------------------------------------------
    Total                 178         40        138       77.53


                 Testing Global Null Hypothesis: BETA=0

                Without          With
  Criterion    Covariates    Covariates     Model Chi-Square

❷ -2 LOG L       123.099        81.227       41.873 with 7 DF (p=0.0001)
  Score             .             .          32.872 with 7 DF (p=0.0001)
  Wald              .             .          22.211 with 7 DF (p=0.0023)


               Analysis of Maximum Likelihood Estimates
                   ❸             ❹            ❺            ❻           ❼
               Parameter     Standard       Wald          Pr >        Risk
  Variable DF   Estimate       Error     Chi-Square    Chi-Square     Ratio

  CHK       1    0.868465     0.62046      1.95920       0.1616       2.383
  AGP1      1    0.137595     0.06372      4.66235       0.0308       1.148
  AGMN      1    0.404447     0.14117      8.20756       0.0042       1.498
  NLV       1    0.379954     0.24648      2.37637       0.1232       1.462
  LIV       1    0.189852     0.15304      1.53890       0.2148       1.209
  WT        1   -0.027759     0.01192      5.42640       0.0198       0.973
  AGLP      1    0.109986     0.06000      3.35978       0.0668       1.116
```

Explanation

This analysis stratifies, or matches, on the age of the subject at the interview (AGMT) and results in N:M matching. Rather than matching each case with a fixed number of controls, varying numbers of cases are matched with varying numbers of controls based on the variable AGMT.

❶ **Summary of the Number of Events and Censored Values** shows that there are 178 observations with usable (nonmissing) data, with 40 cases and 138 controls. Cases have events equal to 1, and controls are coded as censored.

As it turns out, some controls are no longer matched with cases. For example, there are four controls under the age of 31 that now form three strata in which there are no cases (events). Controls such as these do not contribute to this analysis.

❷ **-2 LOG L** gives −2 times the log likelihood function for the null model (without risk factors), the full model (with risk factors), and the likelihood ratio chi-square statistic. The extremely small *p*-value for the chi-square statistic ($p = 0.0001$) indicates that at least one risk factor has a nonzero parameter.

❸ **Parameter Estimate** gives the estimated parameters associated with each risk factor. A positive parameter estimate indicates that the likelihood of benign breast disease increases with the presence, or level, of the risk factor, while a negative parameter estimate indicates that the likelihood decreases with the presence of the risk factor.

❹ **Standard Error** is the standard error of the parameter estimate.

❺ **Wald Chi-Square** is the Wald statistic for the significance of the risk factor in the fitted model. It is the square of the ratio of the parameter estimate to its standard error. Use the associated *p*-value to test whether the association between each risk factor and benign breast disease is nonzero.

❻ **`Pr>Chi-Square`** is the *p*-value for the Wald statistic. Values close to zero indicate that the risk factor has a stronger association with the event than can be explained by chance alone.

Notice that AGP1, AGMN, and WT have *p*-values less than 0.05, AGLP has a *p*-value less than 0.10, and all other risk factors have *p*-values larger than 0.10. This means that the data are consistent with AGP1, AGMN, and WT having a strong association with benign breast disease, AGLP having a moderate association with disease, and the other risk factors having little or no association.

❼ **`Risk Ratio`** is the risk ratio for the risk factor. The risks are the anti-logged parameter estimates. They provide an estimate of the expected change in the risk ratio of having benign breast disease versus not having benign breast disease per unit change in the risk factor.

For example, notice that the risk ratio for AGMN is 1.498. This means that the risk for women having benign breast disease increases 1.498 times (that is, a 47.9 percent increase in the odds) for each unit increase in AGMN.

CAUTION!

For a continuous predictor such as AGMN, the exponentiated parameter estimate is the estimate of the change in risk per unit increase in that predictor and is multiplicative in nature, rather than additive. This means that a two-unit increase in AGMN changes the risk by 1.498^2, or 2.244. ■

A Closer Look

Fit the Model

Because the PHREG procedure is designed to analyze survival time data, you need to trick the procedure to perform conditional logistic regression. As with 1:1 matching, the likelihood maximized by PROC PHREG can be made identical to the likelihood for the conditional logistic model by making some changes to the data. To use PROC PHREG for conditional logistic regression, specify two variables on the left-hand side of the equality in the MODEL statement: a variable that contains a time of occurrence of an event (STATUS) and a variable that indicates whether or not an observation is censored (FNDX). List the risk factors on the right-hand side of the equality.

- STATUS is simply a dummy variable, constructed so that the time of the event for cases is less than that for all controls in the same stratum. In this example, all cases have STATUS=1 and all controls have STATUS=2.
- FNDX indicates whether a subject is a case (FNDX=1) or a control (FNDX=0). FNDX(0) causes PROC PHREG to treat the controls as censored.
- For N:M matched data, you must use TIES=DISCRETE to fit the correct model. (For 1:M matched data, the likelihood for conditional logistic regression reduces to that of the Cox model for the continuous time scale. This means that for 1:M matched data, you can use the default TIES=BRESLOW on the MODEL statement or use TIES=DISCRETE.)

□ The conditional logistic regression is a stratified analysis, where each matched set is a stratum. Use the STRATA statement to specify the matching variable, which is AGMT in this example.

Variation

You may have some reasons to analyze only the subset of matched sets that have complete data for all risk factors. For these data, this would mean discarding 21 matched sets, or a little over 40 percent of the data. This would obviously be too great a price to pay for these data. However, to illustrate this approach, the following example shows how to create the subset of N=29 matched sets and then analyze it with PROC PHREG.

Create a variable, MISSING, to indicate whether or not a subject has complete data for the risk factors.

```
data match_NM;
   set match_NM;
   missing = nmiss( of chk-aglp );
run;
```

Use PROC SQL to subset the data. Create a data set, SUBSET, by keeping only matched sets that have complete data for all risk factors.

```
proc sql;
create table subset (drop=count1) as
   select *, count(matchset) as count1
   from match_NM
   where missing = 0
   group by matchset
   having count(*) = 4
   order by matchset;
```

Fit the conditional logistic regression model. DATA=SUBSET uses the subset of N=29 matched sets.

```
proc phreg data=subset;
   model status*fndx(0) = chk agp1 agmn nlv liv wt aglp;
   strata matchset;
run;
```

Output 14.2
Output from PROC PHREG Using Only Matched Sets Having Complete Data

```
                        The PHREG Procedure

      Data Set: WORK.SUBSET
      Dependent Variable: STATUS
      Censoring Variable: FNDX
      Censoring Value(s): 0
      Ties Handling: BRESLOW

             Summary of the Number of Event and Censored Values

                                                              Percent
       Stratum    MATCHSET        Total       Event   Censored   Censored

             1    1                   4           1          3      75.00
             2    3                   4           1          3      75.00
             3    4                   4           1          3      75.00
             4    6                   4           1          3      75.00
             5    9                   4           1          3      75.00
             6    10                  4           1          3      75.00
             7    12                  4           1          3      75.00
             8    14                  4           1          3      75.00
             9    15                  4           1          3      75.00
            10    16                  4           1          3      75.00
            11    21                  4           1          3      75.00
            12    22                  4           1          3      75.00
            13    23                  4           1          3      75.00
            14    24                  4           1          3      75.00
            15    25                  4           1          3      75.00
            16    30                  4           1          3      75.00
            17    31                  4           1          3      75.00
            18    34                  4           1          3      75.00
            19    35                  4           1          3      75.00
```

20	36	4	1	3	75.00
21	37	4	1	3	75.00
22	38	4	1	3	75.00
23	40	4	1	3	75.00
24	41	4	1	3	75.00
25	43	4	1	3	75.00
26	44	4	1	3	75.00
27	47	4	1	3	75.00
28	48	4	1	3	75.00
29	49	4	1	3	75.00
Total		116	29	87	75.00

Testing Global Null Hypothesis: BETA=0

Criterion	Without Covariates	With Covariates	Model Chi-Square
-2 LOG L	80.405	52.964	27.441 with 7 DF (p=0.0003)
Score	.	.	22.858 with 7 DF (p=0.0018)
Wald	.	.	16.025 with 7 DF (p=0.0249)

Analysis of Maximum Likelihood Estimates

Variable	DF	Parameter Estimate	Standard Error	Wald Chi-Square	Pr > Chi-Square	Risk Ratio
CHK	1	-0.196808	0.70156	0.07870	0.7791	0.821
AGP1	1	0.147246	0.08519	2.98716	0.0839	1.159
AGMN	1	0.468455	0.17173	7.44159	0.0064	1.598
NLV	1	0.374580	0.28451	1.73341	0.1880	1.454
LIV	1	0.072676	0.19955	0.13264	0.7157	1.075
WT	1	-0.014216	0.01234	1.32649	0.2494	0.986
AGLP	1	0.097036	0.06440	2.27039	0.1319	1.102

From the output, you see that the data are consistent with only AGMN having a strong association with disease, AGP1 having a moderate association with disease, and all other risk factors having little or no association. By discarding 40 percent of the data, you have sacrificed a great deal of power to detect associations between risk factors and disease.

EXAMPLE 15

Fitting Interactions in Logistic Regression Models

Featured Tools:

- PROC GENMOD
- PROC LOGISTIC
- DATA step programming

Some logistic regression models require interaction terms to provide a better fit to the data than that provided by simple main effects. When the effect of one explanatory variable on the response depends on the levels of one or more other explanatory variables, then your model may require an interaction term.

When you want to include interaction terms in a model that PROC LOGISTIC fits, you must create the interaction terms in a DATA step before invoking PROC LOGISTIC. You can include interactions for logistic regression models directly, without preliminary DATA step programming, when you use the GENMOD procedure. PROC GENMOD fits generalized linear models, of which logistic regression is one type. PROC GENMOD enables you to specify interaction terms in the MODEL statement by combining two or more variables with an asterisk (for example, A*B). To perform logistic regression with PROC GENMOD, your data must be in the form of a binomial experiment. That is, you must use the *events/trials* model syntax.

PROC GENMOD does not compute odds ratios from its parameter estimates. If you want odds ratios, you must compute them in a DATA step. Use an output data set from PROC GENMOD that contains the parameter estimates and exponentiate the parameter estimates to compute the odds ratios.

This example shows how to perform the following logistic regression tasks:

- fit a simple logistic regression model with PROC GENMOD
- compute odds ratios from the parameter estimates that PROC GENMOD generates
- include interactions with PROC GENMOD
- use DATA step programming so that you can include interactions with PROC LOGISTIC and produce results that exactly match the results that PROC GENMOD produces.

Program

Create the MORTAL data set. The complete MORTAL data set is in the Introduction.

```
data mortal;
   input deaths tbirths cigs age gestpd;
   datalines;
50 365  1 1 1
9 49    2 1 1
41 188  1 2 1
more data lines
;
```

Fit a simple logistic regression model using the GENMOD procedure.

```
proc genmod data=mortal;
```

Fit the model with the MODEL statement. The LINK= option specifies the link function to use in the model. The DIST= option specifies the built-in probability function to use in the model.

```
   model deaths/tbirths=cigs age gestpd / link=logit
                                          dist=binomial;
```

Create an output data set. Use the MAKE statement. The PARMEST table contains an analysis of parameter estimates.

```
   make 'parmest' out=parms;
```

```
   title 'Perinatal Mortality Data';
run;
```

Create odds ratios from the parameter estimates that are contained in the output data set.

```
data parms1;
   set parms;
   if parm='AGE' or parm='CIGS' or parm='GESTPD' then do;
      oddsrat=exp(estimate);
   end;
run;

proc print data=parms1;
run;
```

Include interactions directly by adding interaction terms to the MODEL statement.

```
proc genmod data=mortal;
   model deaths/tbirths = cigs age gestpd
                          age*cigs age*gestpd / link=logit
                                                dist=binomial;
run;
```

Output

Output 15.1
PROC GENMOD Output for a Simple Model

```
                    Perinatal Mortality Data

                     The GENMOD Procedure

                       Model Information

             Description                      Value

             Data Set                         WORK.MORTAL
             Distribution                     BINOMIAL
             Link Function                    LOGIT
             Dependent Variable               DEATHS
             Dependent Variable               TBIRTHS
             Observations Used                8
             Number Of Events                 149
             Number Of Trials                 6751

               Criteria For Assessing Goodness Of Fit

         Criterion              DF          Value       Value/DF

         Deviance                4         1.3694         0.3424
         Scaled Deviance         4         1.3694         0.3424
         Pearson Chi-Square      4         1.3794         0.3449
         Scaled Pearson X2       4         1.3794         0.3449
         Log Likelihood          .      -541.4166              .

                ❶ Analysis Of Parameter Estimates

     Parameter    DF    Estimate     Std Err    ChiSquare  Pr>Chi

     INTERCEPT     1      0.5643      0.4830       1.3648  0.2427
     CIGS          1      0.4162      0.2621       2.5211  0.1123
     AGE           1      0.4866      0.1805       7.2683  0.0070
     GESTPD        1     -3.2878      0.1847     316.8595  0.0001
     SCALE         0      1.0000      0.0000            .       .

NOTE:  The scale parameter was held fixed.
```

Output 15.2
Listing of Revised Output Data Set from PROC GENMOD Showing Odds Ratios

```
                     Perinatal Mortality Data

OBS  PARM       DF   ESTIMATE     STDERR       CHISQ     PVAL   ODDSRAT
                                                                   ❷
 1   INTERCEPT   1     0.5643     0.4830      1.3648   0.2427    .
 2   CIGS        1     0.4162     0.2621      2.5211   0.1123   1.51625
 3   AGE         1     0.4866     0.1805      7.2683   0.0070   1.62678
 4   GESTPD      1    -3.2878     0.1847    316.8595   0.0001   0.03734
 5   SCALE       0     1.0000     0.0000       .        .        .
```

Output 15.3
PROC GENMOD Output for a Model with Interaction Terms

```
                    Perinatal Mortality Data

                     The GENMOD Procedure

                       Model Information

             Description                      Value

             Data Set                         WORK.MORTAL
             Distribution                     BINOMIAL
             Link Function                    LOGIT
             Dependent Variable               DEATHS
             Dependent Variable               TBIRTHS
             Observations Used                8
             Number Of Events                 149
             Number Of Trials                 6751
```

```
                    Criteria For Assessing Goodness Of Fit

             Criterion              DF          Value        Value/DF

             Deviance                2         0.6228          0.3114
             Scaled Deviance         2         0.6228          0.3114
             Pearson Chi-Square      2         0.6299          0.3150
             Scaled Pearson X2       2         0.6299          0.3150
             Log Likelihood          .      -541.0432               .

                      Analysis Of Parameter Estimates

         Parameter     DF    Estimate      Std Err    ChiSquare   Pr>Chi

         INTERCEPT      1     -0.3816       1.1987       0.1013   0.7502
         CIGS           1      0.8996       0.7985       1.2690   0.2599
         AGE            1      1.1968       0.8516       1.9750   0.1599
         GESTPD         1     -2.9851       0.5564      28.7855   0.0001
         CIGS*AGE       1     -0.3765       0.5997       0.3942   0.5301
         AGE*GESTPD     1     -0.2203       0.3864       0.3252   0.5685
         SCALE          0      1.0000       0.0000            .        .

NOTE:  The scale parameter was held fixed.
```

Explanation

Output 15.1 shows the results from fitting a binary logistic regression model with PROC GENMOD. Examine the **Analysis of Parameter Estimates** table ❶ and note that these results are identical to those obtained from PROC LOGISTIC in Output 10.1 (see Example 10).

Output 15.2 shows the output data set of parameter estimates from PROC GENMOD after you compute odds ratios from the estimates. The variable, ODDSRAT, contains the odds ratios for the explanatory variables ❷. These odds ratios are identical to those that PROC LOGISTIC computes in Output 10.1.

Output 15.3 shows the PROC GENMOD output for the model with interaction terms. The interactions are added to demonstrate the technique and show the resulting output, even though these data do not necessarily require interaction terms.

Variation

Including Interactions with PROC LOGISTIC

You can include interactions in your logistic regression model with PROC LOGISTIC, but it requires some preliminary DATA step programming. The following code sample shows how to produce exactly the same results for PROC LOGISTIC and PROC GENMOD for the interactions you include in the model in this example.

Create the interaction terms in a DATA step. AGEXCIGS is the interaction between mother's age and smoking status. AGEXGEST is the interaction between mother's age and the gestation period length.

```
data mortal1;
   set mortal;
   agexcigs=age*cigs;
   agexgest=age*gestpd;
run;
```

Fit the interaction model by adding the interaction terms to the MODEL statement.

```
proc logistic data=mortal1;
   model deaths/tbirths = cigs age gestpd agexcigs agexgest;
run;
```

Output 15.4
Including Interactions with PROC LOGISTIC

```
                         Perinatal Mortality Data

                         The LOGISTIC Procedure

Data Set: WORK.MORTAL1
Response Variable (Events): DEATHS
Response Variable (Trials): TBIRTHS
Number of Observations: 8
Link Function: Logit

                            Response Profile

                        Ordered  Binary
                          Value  Outcome        Count

                              1  EVENT            149
                              2  NO EVENT        6602

                 Testing Global Null Hypothesis: BETA=0

                              Intercept
                 Intercept       and
Criterion          Only       Covariates    Chi-Square for Covariates

AIC              1433.110      1094.086          .
SC               1439.927      1134.991          .
-2 LOG L         1431.110      1082.086     349.023 with 5 DF (p=0.0001)
Score                .             .        717.554 with 5 DF (p=0.0001)

              ❸ Analysis of Maximum Likelihood Estimates

                Parameter Standard    Wald       Pr >     Standardized      Odds
Variable DF  Estimate   Error  Chi-Square Chi-Square   Estimate        Ratio

INTERCPT 1    -0.3816   1.1987     0.1013     0.7502            .          .
CIGS     1     0.8996   0.7985     1.2690     0.2599     0.146709      2.459
AGE      1     1.1968   0.8516     1.9750     0.1599     0.293633      3.310
GESTPD   1    -2.9851   0.5564    28.7855     0.0001    -0.474287      0.051
AGEXCIGS 1    -0.3765   0.5997     0.3942     0.5301    -0.124956      0.686
AGEXGEST 1    -0.2203   0.3864     0.3252     0.5685    -0.112031      0.802

        Association of Predicted Probabilities and Observed Responses

                 Concordant = 76.0%          Somers' D = 0.665
                 Discordant =  9.4%          Gamma     = 0.779
                 Tied       = 14.6%          Tau-a     = 0.029
                 (983698 pairs)              c         = 0.833
```

Explanation

Examine the results for PROC LOGISTIC in the **Analysis of Maximum Likelihood Estimates** table ❸ shown in Output 15.4. Note that those results are identical to the results for PROC GENMOD in the **Analysis of Parameter Estimates** table shown in Output 15.3.

Further Reading

- For complete reference and usage information on the GENMOD procedure, see SAS Technical Report P-243, *SAS/STAT Software: The GENMOD Procedure, Release 6.09.*

EXAMPLE 16

Estimating Discrete Choice Probabilities with a Multinomial Logit Model

Featured Tools:

- PROC PHREG
- DATA step programming

The multinomial logit model, also known as McFadden's model or the conditional logit model (Agresti 1990; McFadden 1974), is a useful tool for investigating consumer choice behavior. It is an alternative to conjoint analysis as a marketing research tool. The PHREG procedure fits multinomial logit, discrete choice models as a special case of survival analysis.

In a discrete choice study, subjects are presented with alternatives and asked to choose the most preferred alternative. In some cases, the set of alternatives may differ from subject to subject, but in this example the set of alternatives is the same for all subjects. The multinomial logit model in this example uses characteristics of the choices as explanatory variables, but multinomial logit models can also use characteristics of the choosers as explanatory variables.

The multinomial logit model is a conditional model conditioned on the set of choices. The model assumes that the probability that an individual will choose one of the m alternatives, c_i, from the set of choices, C, is

$$\text{Prob}(c_i \mid \text{C}) = \frac{\exp[\text{U}(c_i)]}{\sum_{j=1}^{m} \exp[\text{U}(c_j)]} = \frac{\exp(\mathbf{x}_i \beta)}{\sum_{j=1}^{m} \exp(\mathbf{x}_j \beta)}$$

where

$\text{U}(c_i) = \mathbf{x}_i \beta$	is the utility for alternative c_i.
$\mathbf{x}_i$	is a vector of alternative attributes.
β	is a vector of unknown parameters.

In this example, you use PROC PHREG to estimate the parameters and you use the DATA step to compute the estimated probabilities that an attribute combination will be chosen. Because PROC PHREG is designed to analyze survival time data, most data must be rearranged in a form that is consistent with survival data. To make discrete choice data consistent with survival analysis data, the most preferred alternative is said to occur at time one. All other alternatives are said to occur at later times. These other (nonpreferred) alternatives are said to be *censored.*

Program

```
   /* global title statement */
title 'Chocolate Candy Data';
```

Create the CHOCS data set. The complete CHOCS data set is in the Introduction.

```
data chocs;
   input subj choose dark soft nuts @@;
```

Create a dummy time variable, T, so that the value of T for chosen items is smaller than that for nonchosen items. The value of CHOOSE is 1 for the chosen attribute combination, which represents the event to be modeled. The value of CHOOSE is 2 for the other seven nonchosen attribute combinations, which represent nonevents.

```
   t=2-choose;

   datalines;
1 0 0 0 0   1 0 0 0 1  1 0 0 1 0  1 0 0 1 1
1 1 1 0 0   1 0 1 0 1  1 0 1 1 0  1 0 1 1 1
2 0 0 0 0   2 0 0 0 1  2 0 0 1 0  2 0 0 1 1
more data lines
;
```

Fit the multinomial logit model and produce a summary table with PROC PHREG. Create an output data set that contains the estimated logistic regression coefficients. Use the OUTEST= option.

```
proc phreg data=chocs outest=betas;
```

Perform a stratified analysis with each subject as one stratum. Use the STRATA statement to specify the variables that determine the stratification.

```
   strata subj;
```

Fit the model with the dummy time variable, T, as the response and CHOOSE as the censoring variable. Specify within parentheses the censored values of the censoring variable, CHOOSE.

```
   model t*choose(0)=dark soft nuts;
run;
```

Create a data set that contains all the attribute combinations.

```
data combos;
   set chocs;
   if subj=1;
   keep dark soft nuts;
run;
```

Create a data set that merges the attribute combinations with the estimated parameters. Compute the predicted probabilities in the DATA step.

```
data probs;
   retain sumxbeta 0;
   set combos end=eof;
   if _n_=1 then
      set betas(rename=(dark=b1 soft=b2 nuts=b3));
   keep dark soft nuts xbeta;
   array x[3] dark soft nuts;
   array b[3] b1-b3;
```

Compute the dot, or inner, product of the estimated parameters and the attribute combinations.

```
   xbeta=0;
   do j=1 to 3;
      xbeta=xbeta + x[j]*b[j];
   end;
```

Exponentiate the products and sum them.

```
   xbeta=exp(xbeta);
   sumxbeta=sumxbeta+xbeta;
```

Output the summed exponentiated products.

```
   if eof then call symput('sumxbeta',put(sumxbeta,best12.));
run;
```

Format the output.

```
proc format;
   value df 1='dark' 0='milk';
   value sf 1='soft' 0='hard';
   value nf 1='nuts' 0='no nuts';
run;
```

Divide each exponentiated product by the sum of all exponentiated products.

```
data probs1;
   set probs;
   p=xbeta / &sumxbeta;
   drop xbeta;
   format dark df. soft sf. nuts nf.;
run;
```

Sort the probabilities in descending order.

```
proc sort data=probs1;
   by descending p;
run;
```

Print the estimated probabilities.

```
proc print data=probs1;
run;
```

Output

Output 16.1
PROC PHREG Output: Summary Table, Chi-Square Statistics, and Parameter Estimates

```
                         Chocolate Candy Data

                         The PHREG Procedure

     Data Set: WORK.CHOCS
     Dependent Variable: T
     Censoring Variable: CHOOSE
     Censoring Value(s): 0
     Ties Handling: BRESLOW

           Summary of the Number of Event and Censored Values

                                                              Percent
      Stratum     SUBJ          Total       Event   Censored   Censored

            1     1                 8           1          7      87.50
            2     2                 8           1          7      87.50
            3     3                 8           1          7      87.50
            4     4                 8           1          7      87.50
            5     5                 8           1          7      87.50
            6     6                 8           1          7      87.50
            7     7                 8           1          7      87.50
            8     8                 8           1          7      87.50
            9     9                 8           1          7      87.50
           10    10                 8           1          7      87.50
      -------------------------------------------------------------------
        Total                      80          10         70      87.50

              Testing Global Null Hypothesis: BETA=0

                  Without       With
     Criterion    Covariates    Covariates    Model Chi-Square

     -2 LOG L         41.589        28.727      12.862 with 3 DF (p=0.0049)
     Score              .             .         11.600 with 3 DF (p=0.0089)
     Wald               .             .          8.928 with 3 DF (p=0.0303)


             Analysis of Maximum Likelihood Estimates

                      Parameter    Standard      Wald         Pr >          Risk
     Variable  DF     Estimate       Error    Chi-Square   Chi-Square      Ratio

     DARK       1     1.386294     0.79057     3.07490      0.0795        4.000
     SOFT       1    -2.197225 ❷  1.05409     4.34502      0.0371 ❶     0.111
     NUTS       1     0.847298     0.69007     1.50762      0.2195        2.333
```

Output 16.2
Computed Probabilities from Multinomial Logit Model

```
                         Chocolate Candy Data

               OBS    DARK    SOFT    NUTS          P

                1     dark    hard    nuts       0.504 ❸
                2     dark    hard    no nuts    0.216
                3     milk    hard    nuts       0.126
                4     dark    soft    nuts       0.056
                5     milk    hard    no nuts    0.054
                6     dark    soft    no nuts    0.024
                7     milk    soft    nuts       0.014
                8     milk    soft    no nuts    0.006
```

Explanation

Output 16.1 shows the PROC PHREG output from the multinomial logit model. Use the **`Summary of the Number of Event and Censored Values`** table to check your data entry. There must be as many strata as subjects, and each stratum must consist of the *m* observations for one of the subjects. For each stratum, the **`Total`** frequency is *m*, the **`Event`** frequency is 1, and the **`Censored`** frequency is $m - 1$ when the data are arrayed correctly.

In the **`Analysis of Maximum Likelihood Estimates`** table, the parameter estimate with the smallest *p*-value is **`SOFT`**, with a *p*-value of 0.0371 ❶. It has a negative parameter estimate of -2.197 ❷, which means that a unit increase (from 0=hard to 1=soft) decreases preference. This means that hard centers are preferred over soft centers. From the other parameter estimates, you can infer that dark chocolate is preferred over milk chocolate, and nuts are preferred over no nuts. In the latter two cases, however, the *p*-values are not significant at the 0.05 significance level.

The parameter estimates that are contained in the OUTEST= data set are used to construct the estimated probabilities that each alternative would be chosen. Output 16.2 shows the listing of the data set that contains these estimated choice probabilities. The most preferred type of candy consists of dark chocolate with a hard center and nuts. Note that five of the ten subjects chose this type of candy as their most preferred attribute combination. The estimated choice probability for this attribute combination is 0.504 ❸.

Further Reading

- □ For complete reference and usage information on the PHREG procedure, see SAS Technical Report P-229, *SAS/STAT Software: Changes and Enhancements, Release 6.07.*
- □ For more information on DATA step processing, see *SAS Language: Reference, Version 6, First Edition.*

References

- □ Agresti, A. (1990), *Categorical Data Analysis,* New York: John Wiley & Sons, Inc.
- □ McFadden, D. (1974), "Conditional Logit Analysis of Qualitative Choice Behavior," in *Frontiers in Economics,* ed. P. Zarembka, New York: Academic Press, 105-142.

E X A M P L E 17

Computing Likelihoods for Discrete Choice Data with a Multinomial Logit Model

Featured Tools:

- PROC PHREG
- DATA step programming

In a brand choice version of a discrete choice study, subjects are presented with several choice sets consisting of different brands with different prices for the competing brands in each set. The subjects choose one preferred brand from each set. A subject's choice of preferred brand may differ from set to set, depending on the price for each brand in each set.

Consider the BRANDS data set from the Introduction. In those data, five brands are arranged in eight choice sets. Let x_0 be the price of the brand. Let x_1, x_2, x_3, and x_4 be indicator variables representing the choice of brands, such that

$$x_i = 1 \quad \text{if brand } i \text{ is chosen}$$
$$= 0 \quad \text{otherwise}$$

Let $\mathbf{x} = (x_0, x_1, x_2, x_3, x_4)$ be the vector of alternative attributes. Consider the first choice set. There are five distinct vectors of alternative attributes:

$$\mathbf{x}_{11} = (5.99\ 1\ 0\ 0\ 0) \quad \mathbf{x}_{12} = (5.99\ 0\ 1\ 0\ 0) \quad \mathbf{x}_{13} = (5.99\ 0\ 0\ 1\ 0)$$
$$\mathbf{x}_{14} = (5.99\ 0\ 0\ 0\ 1) \quad \mathbf{x}_{15} = (4.99\ 0\ 0\ 0\ 0)$$

The vector $\mathbf{x}_{12}$, for example, represents the choice of Brand 2 at a price of \$5.99. The vector $\mathbf{x}_{15}$ represents the choice of Other at a price of \$4.99.

This example shows how to fit brand choice models, compute likelihoods, and compare different models. The PHREG procedure maximizes the Breslow likelihood. For brand choice models, the choice likelihood is more commonly used. The Breslow likelihood is related to the choice likelihood L_C by a multiplicative constant, as shown in the following formula:

$$L_C = N^{kN} L_B$$

The choice likelihood, L_k^C, for each choice set, k, is expressed in the following formula:

$$L_k^C = \frac{\exp\left[\left(\sum_{j=1}^{m} f_{kj}\mathbf{x}_{kj}\right)\beta\right]}{\left[\sum_{j=1}^{m} \exp\left(\mathbf{x}_{kj}\beta\right)\right]^N}$$

where

m	is the total number of brands in each choice set.
f_{kj}	is the frequency that brand j is chosen in choice set k.
$\mathbf{x}_{kj}$	is the choice vector for brand j in choice set k.
β	is the vector of unknown model parameters.
N	is the total frequency for each choice set.

The likelihood for all q choice sets is the product of the individual likelihoods, as expressed by the following formula:

$$L_C = \Pi_{k=1}^{q} L_k^C$$

This example uses the DATA step to create a design matrix for brand choice data. PROC PHREG then fits a multinomial logit model to the data. The example in "Variation" shows how you can use the DATA step to combine output data sets from PROC PHREG and compute a chi-square test for comparing two models.

Program

Create the BRANDS data set and the design matrix for the analysis. The complete BRANDS data set is in the Introduction.

```
   /* global title statement */
title 'Brand Choice Data';

data brands;
   drop p1-p5 f1-f5 j;
   input p1-p5 f1-f5;
```

Create arrays for brand prices, for brand choice frequencies, and for creating the design matrix.

```
   array p[5] p1-p5;
   array f[5] f1-f5;
   array pb[5] price1-price5;
   array brand[5] brand1-brand5;
```

Initialize the brand and brand-by-price design matrices to 0.

```
   do j=1 to 5;
      brand[j]=0;
      pb[j]   =0;
   end;
```

Count the total number of choices.

```
   nobs=sum(of f1-f5);
```

Store the choice set number to stratify.

```
   ch_set=_n_;
```

Create the design matrix.

```
   do j=1 to 5;
      price    = p[j];
      brand[j] = 1;
      pb[j]    = price;
```

Output the number of times each brand was chosen.

```
      freq   = f[j];
      choose = 1;
      t      = 1;          /* choice occurs at time 1 */
      output;
```

Output the number of times each brand was not chosen.

```
      freq   = nobs-f[j];
      choose = 0;
      t      = 2;          /* nonchoice occurs at time 2 */
      output;
```

Set up for the next alternative.

```
      brand[j] = 0;
      pb[j]    = 0;
   end;

   datalines;
5.99 5.99 5.99 5.99 4.99  12 19 22 33 14
5.99 5.99 3.99 3.99 4.99  34 26  8 27  5
5.99 3.99 5.99 3.99 4.99  13 37 15 27  8
more data lines
;
```

Print the first 20 observations of the brand choice data and design matrix.

```
proc print data=brands(obs=20);
run;
```

Fit the multinomial logit, discrete choice model with a common price slope. Use PROC PHREG with the NOSUMMARY option (to conserve space).

```
proc phreg data=brands nosummary;
   title2 'Discrete Choice with Common Price Effect';
```

Perform a stratified analysis with each choice set as one stratum.

```
   strata ch_set;
```

Fit the model with the dummy time variable, T, as the response and CHOOSE as the censoring variable. Specify within parentheses the censored values of the censoring variable, CHOOSE.

```
   model t*choose(0)=brand1-brand5 price;
```

Specify the frequency variable.

```
   freq freq;
run;
```

Fit the multinomial logit model with brand by price effects. Include all PRICE variables in the MODEL statement.

```
proc phreg data=brands nosummary;
   title2 'Discrete Choice with Brand by Price Effects';
   strata ch_set;
   model t*choose(0)=brand1-brand5 price1-price5;
   freq freq;
run;
```

Output

Output: PROC PRINT

Output 17.1
Partial Listing of Brand Choice Data and Design Matrix

Brand Choice Data

OBS	PRICE1	PRICE2	PRICE3	PRICE4	PRICE5	BRAND1	BRAND2	BRAND3	BRAND4	BRAND5	NOBS	CH_SET	PRICE	FREQ	CHOOSE	T
1	5.99	0.00	0.00	0.00	0.00	1	0	0	0	0	100	1	5.99	12	1	1
2	5.99	0.00	0.00	0.00	0.00	1	0	0	0	0	100	1	5.99	88	0	2
3	0.00	5.99	0.00	0.00	0.00	0	1	0	0	0	100	1	5.99	19	1	1
4	0.00	5.99	0.00	0.00	0.00	0	1	0	0	0	100	1	5.99	81	0	2
5	0.00	0.00	5.99	0.00	0.00	0	0	1	0	0	100	1	5.99	22	1	1
6	0.00	0.00	5.99	0.00	0.00	0	0	1	0	0	100	1	5.99	78	0	2
7	0.00	0.00	0.00	5.99	0.00	0	0	0	1	0	100	1	5.99	33	1	1
8	0.00	0.00	0.00	5.99	0.00	0	0	0	1	0	100	1	5.99	67	0	2
9	0.00	0.00	0.00	0.00	4.99	0	0	0	0	1	100	1	4.99	14	1	1
10	0.00	0.00	0.00	0.00	4.99	0	0	0	0	1	100	1	4.99	86	0	2
11	5.99	0.00	0.00	0.00	0.00	1	0	0	0	0	100	2	5.99	34	1	1
12	5.99	0.00	0.00	0.00	0.00	1	0	0	0	0	100	2	5.99	66	0	2
13	0.00	5.99	0.00	0.00	0.00	0	1	0	0	0	100	2	5.99	26	1	1
14	0.00	5.99	0.00	0.00	0.00	0	1	0	0	0	100	2	5.99	74	0	2
15	0.00	0.00	3.99	0.00	0.00	0	0	1	0	0	100	2	3.99	8	1	1
16	0.00	0.00	3.99	0.00	0.00	0	0	1	0	0	100	2	3.99	92	0	2
17	0.00	0.00	0.00	3.99	0.00	0	0	0	1	0	100	2	3.99	27	1	1
18	0.00	0.00	0.00	3.99	0.00	0	0	0	1	0	100	2	3.99	73	0	2
19	0.00	0.00	0.00	0.00	4.99	0	0	0	0	1	100	2	4.99	5	1	1
20	0.00	0.00	0.00	0.00	4.99	0	0	0	0	1	100	2	4.99	95	0	2

Explanation

Output 17.1 lists the first 20 observations of the revised BRANDS data set. The output shows the structure of the design matrix and how the data need to be arranged to use PROC PHREG to fit the multinomial logit model.

Output: Common Price Slope Model

Output 17.2
PROC PHREG Output from Common-Price-Slope Model

```
                         Brand Choice Data
               Discrete Choice with Common Price Effect

                        The PHREG Procedure

Data Set: WORK.BRANDS
Dependent Variable: T
Censoring Variable: CHOOSE
Censoring Value(s): 0
Frequency Variable: FREQ
Ties Handling: BRESLOW

                        The PHREG Procedure

             Testing Global Null Hypothesis: BETA=0

              Without        With
Criterion    Covariates    Covariates    Model Chi-Square

-2 LOG L      9943.373      9793.486 ❶   149.887 with 5 DF (p=0.0001)
Score             .             .        153.233 with 5 DF (p=0.0001)
Wald              .             .        142.901 with 5 DF (p=0.0001)
```

```
                         The PHREG Procedure

               Analysis of Maximum Likelihood Estimates

                 Parameter    Standard      Wald         Pr >         Risk
Variable  DF     Estimate      Error     Chi-Square  Chi-Square      Ratio

BRAND1     1     0.667270     0.12305     29.40669      0.0001       1.949
BRAND2     1     0.385033     0.12962      8.82360      0.0030       1.470
BRAND3     1    -0.159544     0.14725      1.17399      0.2786       0.853
BRAND4     1     0.989640     0.11720     71.29955      0.0001       2.690
BRAND5     0            0           .            .           .           .
PRICE      1     0.149663     0.04406     11.53792      0.0007       1.161
```

Explanation

In Output 17.2 PROC PHREG displays the −2 LOG L value of 9793.486 ❶, which is $-2\ln(L_B)$ for all q choice sets in the model. Recall that the Breslow likelihood is related to the choice likelihood L_C by a multiplicative constant. For the BRANDS data, $k = 8$, and $N = 100$. You can, therefore, compute the log of the choice likelihood as follows:

$$\log\left(L_C\right) = \left[800 \times \log(100)\right] + \left[(-0.5) \times 9793.486\right]$$

$$\log\left(L_C\right) = -1212.60$$

The parameter estimates shown in Output 17.2 use BRAND5 (Other) as a baseline. PROC PHREG automatically sets the last parameter estimate to 0 because the sum of the design matrix columns for the five brands (BRAND1–BRAND5) is redundant with the intercept. From these results, you conclude that BRAND4 is the most preferred, followed by BRAND1 and BRAND2. BRAND3 has a negative parameter estimate, from which you conclude that it is less preferred than BRAND5 (Other). Note, however, that the parameter estimate for BRAND3 is not statistically significant at the 0.05 significance level.

Output: Brand by Price Effect Model

Output 17.3
PROC PHREG Output from Model with Brand by Price Effects

```
                          Brand Choice Data
               Discrete Choice with Brand by Price Effects

                         The PHREG Procedure

Data Set: WORK.BRANDS
Dependent Variable: T
Censoring Variable: CHOOSE
Censoring Value(s): 0
Frequency Variable: FREQ
Ties Handling: BRESLOW


                         The PHREG Procedure

               Testing Global Null Hypothesis: BETA=0

              Without          With
Criterion   Covariates      Covariates     Model Chi-Square

-2 LOG L      9943.373        9793.084 ❷   150.289 with 8 DF (p=0.0001)
Score             .               .        154.256 with 8 DF (p=0.0001)
Wald              .               .        143.143 with 8 DF (p=0.0001)
```

```
                        The PHREG Procedure

              Analysis of Maximum Likelihood Estimates

                  Parameter    Standard       Wald          Pr >        Risk
Variable   DF     Estimate     Error     Chi-Square    Chi-Square     Ratio

BRAND1      1    -0.009718    0.43555    0.0004978       0.9822       0.990
BRAND2      1    -0.622297    0.48866    1.62172         0.2029       0.537
BRAND3      1    -0.812513    0.60318    1.81451         0.1780       0.444
BRAND4      1     0.317787    0.39549    0.64566         0.4217       1.374
BRAND5      0            0          .          .              .           .
PRICE1      1     0.135870    0.08259    2.70631         0.1000       1.146
PRICE2      1     0.200742    0.09210    4.75120         0.0293       1.222
PRICE3      1     0.131260    0.11487    1.30569         0.2532       1.140
PRICE4      1     0.134778    0.07504    3.22554         0.0725       1.144
PRICE5      0            0          .          .              .           .
```

Explanation

Output 17.3 shows the results of fitting a model with brand by price effects. The −2 LOG L value of 9793.084 for this model ❷ is very similar to that of the common-price-slope model shown in Output 17.2. You can test the null hypothesis that the two models are not significantly different by comparing the likelihoods for the two models. The difference between the two − 2 LOG L values has a chi-square distribution with degrees of freedom equal to the difference between the degrees of freedom for the two likelihoods. Thus, the chi-square test that this model is significantly different from the simpler one with a common price slope is $9793.486 - 9793.084 = 0.402$ with $8 - 5 = 3$ degrees of freedom. Because this chi-square test is not statistically significant, you can conclude that the common-price-slope model fits the data as well as the more complicated brand-by-price-effects model does. See "Variation" for a way to compute this test in a DATA step.

Both the common-price-slope model and the brand-by-price-effects model assume that choices are independent of irrelevant alternatives. For example, subjects' relative preferences for the five brands in this study are unaffected by the addition of another brand to the choice set, according to this assumption. This assumption is often not met in practice.

Variation

Computing the Chi-Square Test in a DATA Step

Use the following code to compute in a DATA step the chi-square test that compares the common-price-slope model to the brand-by-price-effect model.

Create output data sets that contain the log likelihood values for each model. Use the OUTEST= option in the PROC PHREG statement. Use the NOPRINT option to suppress all printed output.

```
proc phreg data=brands noprint outest=stats1;
   strata ch_set;
   model t*choose(0)=brand1-brand5 price;
   freq freq;
run;
```

```
proc phreg data=brands noprint outest=stats2;
   strata ch_set;
   model t*choose(0)=brand1-brand5 price1-price5;
   freq freq;
run;
```

Combine the two OUTEST= data sets. Create a chi-square test statistic by taking the difference between the two log likelihood values and multiplying by -2. Compute the *p*-value with the PROBCHI function.

```
data modtest(keep=chi_sq p_value);
   merge stats1 stats2(rename=(_lnlike_=lnlike2));
   chi_sq = -2*(_lnlike_-lnlike2);
   p_value = 1-probchi(chi_sq,3);
run;

proc print data=modtest;
   title2 'Chi-Square Test for Comparing Models';
run;
```

Output 17.4 shows the results of the chi-square test.

Output 17.4
Computing a Chi-Square Test in the DATA Step

```
                Brand Choice Data
       Chi-Square Test for Comparing Models

            OBS      CHI_SQ      P_VALUE

             1      0.40222      0.93978
```

Explanation

In Output 17.4, the chi-square value of 0.402 has a probability value of 0.94, which means that it is not statistically significant. You cannot reject the null hypothesis that the two models are the same. Thus, you can use the simpler common-price-slope model for these data.

Further Reading

- For complete reference and usage information on the PHREG procedure, see SAS Technical Report P-229, *SAS/STAT Software: Changes and Enhancements, Release 6.07.*
- For more information on DATA step programming, see *SAS Language: Reference, Version 6, First Edition.*

EXAMPLE 18

Probit Analysis for Estimating an LD50

Featured tools:

- DATA step programming
- PROC PROBIT
  - procedure option INVERSECL
  - MODEL statement, events/trials syntax
  - MODEL statement, LACKFIT option
  - OUTPUT statement
- PROC GPLOT, PLOT statement

Toxicity studies are commonly used to investigate the effect a chemical has on an environment and the organisms that inhabit that environment. An *LD50* (lethal dose) is the chemical concentration required to produce a 50 percent mortality rate. In these studies, the response is binary and indicates whether or not the organism died after exposure to a concentration of the chemical. Probit analysis analyzes binary response variables within the linear regression framework.

Let Y be the binary variable that indicates whether an organism survived (Y=1) or died (Y=0) after being exposed to dose level X. Define p as the probability of response, Prob(Y=0). The probit equation is

$$p = \text{Prob}(Y = 0) = \text{C} + (1 - \text{C})\text{F}\big(b_0 + b_1 \times \log(x)\big)$$

where

C	is the natural response rate (that is, the response rate at a zero dose level).
b_0	is an intercept term.
b_1	is the slope coefficient.
x	is the dose. The logarithm of x is usually modeled.
F	is the normal cumulative distribution function.

In this example, you use data from a toxicity study and fit a probit model using the PROBIT procedure. You calculate the quantile estimates and confidence intervals by using the INVERSECL option. You compute a test for goodness of fit with the LACKFIT option. Finally, you create an output data set and plot the probit probability curve with the GPLOT procedure.

Program

Set graphics options.

```
goptions cback=white colors=(black);
```

Create the LD50 data set.

```
data ld50;
   infile datalines eof=endfile;
   input dose n y;
   p_hat=y/n;
   log_dose=log(dose);
   output;
   return;
endfile: do dose=1 to 17 by .5;
            log_dose=log(dose);
            output;
         end;
   datalines;
   0   49   0
 2.6   50   6
 3.8   48  16
 5.1   46  24
 7.7   49  42
10.2   50  44
;
```

Fit the probit model.

```
proc probit data=ld50 log optc;
   model y/n = dose / d=normal inversecl lackfit;
   output out=new p=prob;
   title1 'Probit Analysis for LD50 Data';
run;
```

Reset TITLE1.

```
title1;
```

Sort the data before plotting.

```
proc sort data=new;
   by log_dose;
run;
```

Define two plotting symbols.

```
symbol1 i=none value=diamond height=.75;
symbol2 i=join l=2 value=star height=.75;
```

Plot the probit curve on the logarithmic scale. Overlay the plots of the observed (P_HAT) and the estimated (PROB) response rates. Draw a reference line at 50 percent.

```
proc gplot data=new;
   plot p_hat*log_dose=1 prob*log_dose=2 / frame overlay vref=.5;
   title2 'Probit Plot of Observed and Fitted Probabilities';
run;
```

Output

Output: PROC PROBIT

Output 18.1
Output from PROC PROBIT

```
                              Probit Analysis for LD50 Data

                                    Probit Procedure

               Data Set         =WORK.LD50
               Dependent Variable=Y
               Dependent Variable=N
               Number of Observations=   6
               Number of Events      =      132   Number of Trials =      292
               Observations with Missing Values=  33

      (1)  Log Likelihood for NORMAL -120.3905822

                         (2)  Goodness-of-Fit Tests

                     Statistic                    Value     DF   Prob>Chi-Sq
                     ------------------         --------    --   -----------
                     Pearson Chi-Square          2.4165      3        0.4906
                     L.R.    Chi-Square          2.4169      3        0.4905

                     Response Levels:   2  Number of Covariate Values:    6

NOTE: Since the chi-square is small (p > 0.1000), fiducial limits will be calculated using a t value of  1.96.

                                 (3)          (4)         (5)        (6)
                   Variable  DF   Estimate   Std Err ChiSquare   Pr>Chi Label/Value

                   INTERCPT   1 -2.6884507 0.571302  22.14481  0.0001 Intercept
                   Ln(DOSE)   1 1.74822297 0.276937  39.85043  0.0001
                   C          1          0 0.114596                   Lower threshold

                        Probit Model in Terms of Tolerance Distribution

                                           MU          SIGMA
                                      1.537819      0.572009

                       Estimated Covariance Matrix for Tolerance Parameters

                                         MU               SIGMA                C

                          MU        0.012343           -0.006529         0.011165
                       SIGMA       -0.006529            0.008211        -0.006974
                           C        0.011165           -0.006974         0.013132
                                           Probit Procedure
                                        Probit Analysis on DOSE
        (7)                   (8)          (9)                       (8)            (9)
     Probability          Ln(DOSE) 95 Percent Fiducial Limits              DOSE 95 Percent Fiducial Limits
                                            Lower        Upper

        0.01               0.20713     -0.62119      0.65446        1.23014      0.53731      1.92410
        0.02               0.36306     -0.39727      0.77567        1.43772      0.67215      2.17204
        0.03               0.46199     -0.25542      0.85279        1.58723      0.77459      2.34618
        0.04               0.53641     -0.14884      0.91094        1.70986      0.86170      2.48666
        0.05               0.59695     -0.06225      0.95834        1.81656      0.93965      2.60736
        0.06               0.64847      0.01138      0.99876        1.91262      1.01144      2.71492
        0.07               0.69365      0.07587      1.03427        2.00101      1.07882      2.81305
        0.08               0.73410      0.13355      1.06612        2.08362      1.14288      2.90408
        0.09               0.77089      0.18597      1.09513        2.16170      1.20438      2.98957
        0.10               0.80476      0.23417      1.12189        2.23616      1.26386      3.07064
        0.15               0.94497      0.43319      1.23320        2.57273      1.54218      3.43219
        0.20               1.05640      0.59060      1.32244        2.87601      1.80506      3.75257
        0.25               1.15200      0.72491      1.39972        3.16453      2.06455      4.05408
        0.30               1.23786      0.84481      1.46985        3.44822      2.32753      4.34859
        0.35               1.31741      0.95514      1.53561        3.73375      2.59902      4.64415
        0.40               1.39290      1.05897      1.59887        4.02652      2.88339      4.94742
        0.45               1.46594      1.15843      1.66106        4.33161      3.18493      5.26489
        0.50               1.53782      1.25514      1.72345        4.65443      3.50832      5.60383
        0.55               1.60970      1.35039      1.78729        5.00130      3.85895      5.97323
        0.60               1.68274      1.44536      1.85398        5.38026      4.24337      6.38520
        0.65               1.75823      1.54114      1.92529        5.80214      4.66992      6.85711
        0.70               1.83778      1.63895      2.00356        6.28258      5.14977      7.41542
```

0.75	1.92363	1.74030	2.09224	6.84579	5.69905	8.10302
0.80	2.01923	1.84747	2.19667	7.53256	6.34373	8.99501
0.85	2.13067	1.96470	2.32609	8.42050	7.13274	10.23780
0.90	2.27088	2.10171	2.49941	9.68791	8.18015	12.17528
0.91	2.30474	2.13334	2.54273	10.02161	8.44302	12.71439
0.92	2.34153	2.16716	2.59035	10.39717	8.73343	13.33437
0.93	2.38199	2.20375	2.64329	10.82638	9.05895	14.05933
0.94	2.42716	2.24397	2.70306	11.32672	9.43071	14.92541
0.95	2.47869	2.28911	2.77198	11.92564	9.86611	15.99024
0.96	2.53923	2.34127	2.85380	12.66989	10.39445	17.35365
0.97	2.61365	2.40434	2.95546	13.64879	11.07117	19.21048
0.98	2.71258	2.48674	3.09203	15.06814	12.02205	22.02175
0.99	2.86851	2.61414	3.30976	17.61079	13.65550	27.37856

NOTE: The above quantiles and fiducial limits refer to effects due to the independent variable and do not include any effect due to the natural threshold.

Explanation

❶ `Log Likelihood` is the value of the log likelihood function for the normal distribution.

❷ `Goodness-of-Fit Tests` are a test based on the Pearson chi-square and a test based on the likelihood ratio chi-square. The large *p*-values for these tests indicate that the fitted model agrees with the data very well.

❸ `Estimate` contains the parameter estimates of the probit model. The positive sign of the estimate associated with Ln(DOSE) (the natural logarithm of dose) indicates that the probability of a response (in this case, death) increases with dose. Note that the natural response rate, C, is estimated to be zero. This agrees exactly with the observed data.

❹ `Std Err` is the standard error of the parameter estimate.

❺ `ChiSquare` is an approximate chi-square test that the parameter estimate has no effect on the probability of a response.

❻ `Pr>Chi` is the *p*-value associated with the chi-square test statistic. Both the intercept and the slope for Ln(DOSE) are significant. This indicates that the probability of response does increase significantly as the dose increases.

❼ `Probability` is the probability of a response. The LD50 has a probability equal to 0.50.

❽ `Ln(DOSE)` and `DOSE` give the estimate of the LD50, for the original and the logarithmic scales. A dose level of 4.65 is estimated to produce a 50 percent mortality rate.

❾ `95 Percent Fiducial Limits` gives the lower and upper bounds of a 95 percent confidence interval for the original and the logarithmic scales. The interval for the LD50 is (3.50832, 5.60383).

Output: Probit Curve

Output 18.2
Probit Curve for Observed and Estimated Responses

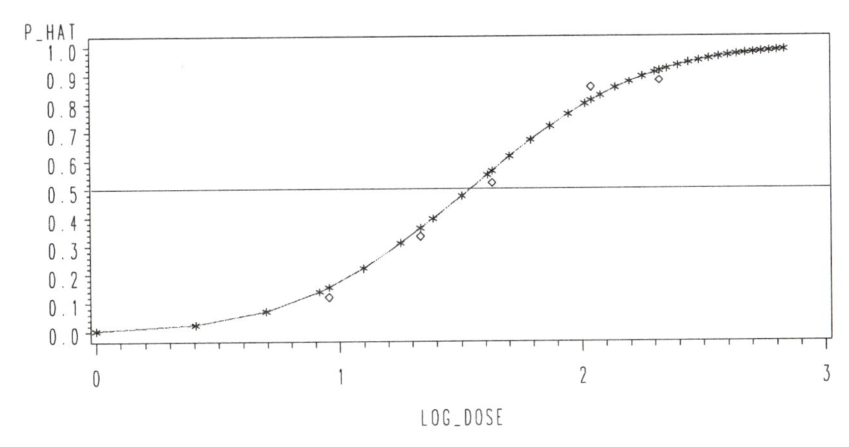

Explanation

The probit curve gives the probability of a response at a given dose level. Typically, the probability approaches 0 for small dose values. For intermediate dose values, the curve increases in an S-curved shape. For large dose values, the probability approaches 1.

A Closer Look

Create the LD50 Data Set

The DATA step that reads the observed data can be used to generate values for the DOSE variable that are used later when plotting the probit curve.

`infile datalines eof=endfile`
: does two things. First, **`datalines`** specifies that the input data immediately follows the DATALINES statement. Second, **`eof=endfile`** jumps to the labeled statement when the end-of-file is reached.

`input dose n y`
: reads in the variables. The variable *N* represents the number of organisms exposed to the corresponding level of the DOSE variable, and *Y* represents the number of organisms that died (or had a positive response) after exposure to the dose.

DO loop
: generates and outputs values of the DOSE variable ranging from 1 to 17 in increments of 0.5. These observations have a missing value for the response variable *Y*. Although PROC PROBIT does not use these observations to fit the model, it does calculate predicted values for them. You use these data points to fill in the plot of the fitted probit curve using PROC GPLOT.

Fit the Probit Model

LOG option on the PROC PROBIT statement
: models the response as a function of the natural logarithm. Use LOG10 when you want to use the base 10 logarithm.

OPTC option with LOG option
: treats the observations with a zero dose level as a control group. The initial estimate of the natural response rate is the ratio of the number of responses to the number of subjects in this group. OPTC is a way to deal with naturally occurring, nonzero response rates. If you know and want to fix the natural response rate at a specific value, use the C= option. Omitting both OPTC and C= assumes a natural response rate of zero.

MODEL Y/N=
: is the correct syntax when your data are in the events/trials format.

D=
: specifies the cumulative distribution function used to model the response. The probit model uses the default, NORMAL.

INVERSECL
: prints the inverse confidence limits.

LACKFIT
: prints goodness-of-fit test statistics.

OUTPUT OUT=NEW P=PROB
: creates an output data set named NEW. NEW contains all variables in the input data set (LD50) plus the fitted probabilities for each observation. You use this data set as an input to PROC GPLOT when you plot the probit curve.

Variation

Fitting Other Models with PROC PROBIT

You can use PROC PROBIT to fit a gompit model or a logit model by using the D= option on the MODEL statement to change the cumulative distribution function used to model the response. The default is the normal distribution. For example,

```
model y/n = x / d=gompertz
```

or

```
model y/n = x / d=logistic
```

EXAMPLE 19

Fitting a Bradley-Terry Model for Paired Comparisons

Featured Tools:

- PROC LOGISTIC
  - OUTEST= option
  - FREQ statement
  - NOINT option
- PROC SORT
- PROC TRANSPOSE
- DATA step programming

Suppose you want to obtain a ranking of consumers' preferences for I items or brands. If the number of items is large, it may be difficult for the consumers to compare all of the items at one time and rank them. However, if the items are compared one pair at a time, the consumers should be able to state a preference for one of the items in the pair. Using the results of all pairwise comparisons, you can establish an overall ranking for all the items.

Bradley and Terry (1952) proposed a logit model for producing rankings from paired comparisons. Let $\beta_1, \beta_2, \ldots, \beta_n$ be regression coefficients associated with the n items $I_1, I_2, \ldots, I_n$, respectively. The probability that I_i is preferred to I_j is

$$\pi_{ij} = \frac{\exp\left(\beta_i\right)}{\exp\left(\beta_i\right) + \exp\left(\beta_j\right)}$$

$$= \frac{\exp\left(\beta_i - \beta_j\right)}{1 + \exp\left(\beta_i - \beta_j\right)}$$

For the qth paired comparison, if I_i is preferred to I_j, let the vector $\boldsymbol{d}_q = \left(d_{q1}, d_{q2}, \ldots, d_{qn}\right)$ be such that

$$d_{qk} = \begin{cases} 1 & k = i \\ -1 & k = j \\ 0 & \text{otherwise} \end{cases}$$

The likelihood for the Bradley-Terry model is identical to the binary logistic model with $\boldsymbol{d}_q$ as explanatory variables, no intercept, and a constant response.

For example, if there are 5 items to compare, then the data set for the analysis requires $5 \times 4 = 20$ observations and seven variables. In each observation, one variable has a value of 1, one variable has a value of -1, and three variables have a value of 0. A frequency variable denotes how many subjects in the study expressed each of the 20 different preferences. A constant response variable is also required.

This example shows how you can use the DATA step to create a data set that contains the necessary covariates. You then use the LOGISTIC procedure to estimate the parameters for the Bradley-Terry model. To obtain the ranking of the items, sort the parameters in descending order.

Program

Create the PAIRS data set that contains the original paired comparison data. The complete PAIRS data set is in the Introduction.

```
   /* global title statement */
title 'Paired Comparison Data';
data pairs;
   input a b c d e f g h;
   datalines;
 .  17 39 44  7 40 18 23
 43 .  25 39 13 17 30 35
 21 35 .  51 11 14 48 26
more data lines
;
```

Revise the original data set to prepare it for the analysis.

```
data pairs2;
   retain i x1-x8 0 resp 1;
   array item{*} a b c d e f g h;
   array x{*} x1-x8;
   set pairs;
   i+1;
   do j=1 to 8;
      if item{j} ne . then do;
```

Create the frequency variable.

```
         count=item{j};
```

Assign values to the covariates.

```
         x{i}=1;
         x{j}=-1;
         output;
         x{i}=0;
         x{j}=0;
      end;
   end;
   drop i j a b c d e f g h;
run;
```

Print the analysis data set.

```
proc print data=pairs2;
   title2 'PAIRS2 Data Set';
run;
```

Fit the Bradley-Terry model. Create an output data set that contains parameter estimates.

```
proc logistic data=pairs2 outest=parms;
```

Fit a model with no intercept. Use the NOINT option in the MODEL statement.

```
   model resp=x1-x8 / noint;
```

Specify a frequency variable.

```
   freq count;
run;
```

Transpose the OUTEST= data set that contains the parameters.

```
proc transpose data=parms out=parms1;
run;
```

Sort the transposed data set by descending values of the parameter estimates.

```
proc sort data=parms1;
   where _name_ ne '_LNLIKE_';
   by descending estimate;
run;
```

Print the sorted parameter estimates to obtain the final ranking of the items.

```
proc print data=parms1;
   title2 'Final Ranking of Items';
run;
```

Output

Output 19.1
Listing of Revised Paired Comparison Data

Paired Comparison Data
PAIRS2 Data Set

OBS	X1	X2	X3	X4	X5	X6	X7	X8	RESP	COUNT
1	1	-1	0	0	0	0	0	0	1	17
2	1	0	-1	0	0	0	0	0	1	39
3	1	0	0	-1	0	0	0	0	1	44
4	1	0	0	0	-1	0	0	0	1	7
5	1	0	0	0	0	-1	0	0	1	40
6	1	0	0	0	0	0	-1	0	1	18
7	1	0	0	0	0	0	0	-1	1	23
8	-1	1	0	0	0	0	0	0	1	43
9	0	1	-1	0	0	0	0	0	1	25
10	0	1	0	-1	0	0	0	0	1	39
11	0	1	0	0	-1	0	0	0	1	13
12	0	1	0	0	0	-1	0	0	1	17
13	0	1	0	0	0	0	-1	0	1	30
14	0	1	0	0	0	0	0	-1	1	35
15	-1	0	1	0	0	0	0	0	1	21
16	0	-1	1	0	0	0	0	0	1	35
17	0	0	1	-1	0	0	0	0	1	51
18	0	0	1	0	-1	0	0	0	1	11
19	0	0	1	0	0	-1	0	0	1	14
20	0	0	1	0	0	0	-1	0	1	48
21	0	0	1	0	0	0	0	-1	1	26
22	-1	0	0	1	0	0	0	0	1	16
23	0	-1	0	1	0	0	0	0	1	21
24	0	0	-1	1	0	0	0	0	1	9
25	0	0	0	1	-1	0	0	0	1	3
26	0	0	0	1	0	-1	0	0	1	21
27	0	0	0	1	0	0	-1	0	1	18
28	0	0	0	1	0	0	0	-1	1	11
29	-1	0	0	0	1	0	0	0	1	53
30	0	-1	0	0	1	0	0	0	1	47
31	0	0	-1	0	1	0	0	0	1	49
32	0	0	0	-1	1	0	0	0	1	57
33	0	0	0	0	1	-1	0	0	1	39
34	0	0	0	0	1	0	-1	0	1	31
35	0	0	0	0	1	0	0	-1	1	58
36	-1	0	0	0	0	1	0	0	1	20
37	0	-1	0	0	0	1	0	0	1	43
38	0	0	-1	0	0	1	0	0	1	46
39	0	0	0	-1	0	1	0	0	1	39
40	0	0	0	0	-1	1	0	0	1	21
41	0	0	0	0	0	1	-1	0	1	40
42	0	0	0	0	0	1	0	-1	1	22
43	-1	0	0	0	0	0	1	0	1	42
44	0	-1	0	0	0	0	1	0	1	30
45	0	0	-1	0	0	0	1	0	1	12
46	0	0	0	-1	0	0	1	0	1	42
47	0	0	0	0	-1	0	1	0	1	29
48	0	0	0	0	0	-1	1	0	1	20
49	0	0	0	0	0	0	1	-1	1	17
50	-1	0	0	0	0	0	0	1	1	37
51	0	-1	0	0	0	0	0	1	1	25
52	0	0	-1	0	0	0	0	1	1	34
53	0	0	0	-1	0	0	0	1	1	49
54	0	0	0	0	-1	0	0	1	1	2
55	0	0	0	0	0	-1	0	1	1	38
56	0	0	0	0	0	0	-1	1	1	43

Output 19.2
PROC LOGISTIC Output for Bradley-Terry Model

```
                         Paired Comparison Data

                         The LOGISTIC Procedure

  Data Set: WORK.PAIRS2
  Response Variable: RESP
  Response Levels: 1
  Number of Observations: 56
  Frequency Variable: COUNT
  Link Function: Logit


                            Response Profile

                        Ordered
                         Value     RESP     Count

                             1        1      1680


                Testing Global Null Hypothesis: BETA=0

                Without        With
 Criterion    Covariates    Covariates    Chi-Square for Covariates

 AIC            2328.975      2078.893            .
 SC             2328.975      2116.879            .
 -2 LOG L       2328.975      2064.893       264.081 with 7 DF (p=0.0001)
 Score              .             .          244.583 with 7 DF (p=0.0001)

NOTE: The following parameters have been set to 0, since the variables are
      a linear combination of other variables as shown.

      X8 = -1 * X1 - 1 * X2 - 1 * X3 - 1 * X4 - 1 * X5 - 1 * X6 - 1 * X7

                Analysis of Maximum Likelihood Estimates

             Parameter Standard    Wald       Pr >    Standardized     Odds
 Variable DF  Estimate   Error  Chi-Square Chi-Square   Estimate       Ratio
                 ❶
 X1        1   -0.3599  0.1348     7.1305     0.0076    -0.099091     0.698
 X2        1   -0.2338  0.1344     3.0271     0.0819    -0.064451     0.792
 X3        1   -0.1979  0.1343     2.1703     0.1407    -0.054557     0.820
 X4        1   -1.2334  0.1462    71.1504     0.0001    -0.328019     0.291
 X5        1    1.0687  0.1491    51.3651     0.0001     0.281566     2.912
 X6        1    0.0271  0.1345     0.0407     0.8401     0.007472     1.028
 X7        1   -0.3237  0.1346     5.7837     0.0162    -0.089189     0.723
 X8        0         0 ❷    .          .          .            .         .

NOTE: Measures of association between the observed and predicted values
      were not calculated because the predicted probabilities are
      indistinguishable when they are classified into intervals of length
      0.002 .
```

Output 19.3
Listing of Sorted Parameter Estimates

```
                         Paired Comparison Data
                         Final Ranking of Items

                        OBS    _NAME_    ESTIMATE

                         1       X5       1.06874
                         2       X6       0.02713
                         3       X8       0.00000
                         4       X3      -0.19786
                         5       X2      -0.23378
                         6       X7      -0.32374
                         7       X1      -0.35985
                         8       X4      -1.23342
```

Explanation

Output 19.1 lists the data set that you use in this analysis. It shows the structure of the data set that PROC LOGISTIC requires to fit the Bradley-Terry model to paired comparison data.

Output 19.2 shows the results of fitting a logistic regression model to the revised paired comparison data. There are eight parameter estimates ❶ that represent the eight items that were compared in the study. The model is parameterized so that the eighth item is assigned a parameter estimate of 0 ❷. The other parameter estimates use the eighth estimate as a reference. Thus, positive parameter estimates indicate items that are preferred to the eighth item, and negative estimates indicate items that are less preferred than the eighth item.

Output 19.3 shows a listing of the final rankings. The fifth item has the highest ranking, and the fourth item has the lowest ranking of the eight items.

Further Reading

- For complete reference information on the SORT and TRANSPOSE procedures, see the *SAS Procedures Guide, Version 6, Third Edition.*
- For more information on DATA step programming, see *SAS Language: Reference, Version 6, First Edition.*

Reference

Bradley, R.A. and Terry, M.E. (1952), “Rank Analysis of Incomplete Block Designs: I. The Method of Paired Comparison,” *Biometrika*, 39, 324-345.

Index

D

E

F

G

H

I

J

L

P

T

U

W

Your Turn

If you have comments or suggestions about *Logistic Regression Examples Using the SAS® System, Version 6, First Edition*, please send them to us on a photocopy of this page or send us electronic mail.

For comments about this book, please return the photocopy to

SAS Institute Inc.
Publications Division
SAS Campus Drive
Cary, NC 27513
email: yourturn@unx.sas.com

For suggestions about the software, please return the photocopy to

SAS Institute Inc.
Technical Support Division
SAS Campus Drive
Cary, NC 27513
email: suggest@unx.sas.com